Chromosome Microdissection and Cloning
A Practical Guide

Chromosome Microdissection and Cloning
A Practical Guide

Nabil G. Hagag and Michael V. Viola
Oncology Division
Department of Medicine
State University of New York at Stony Brook
Stony Brook, New York

Academic Press, Inc.
A Division of Harcourt Brace & Company
San Diego   New York   Boston   London   Sydney   Tokyo   Toronto

Front cover photograph: Microdissection of a metaphase chromosome from normal human peripheral blood lymphocytes. The chromosomes in the spread were stained with a mixture of hematoxylin and propidium iodide and the microdissection needle was filled with lucifer yellow. Photomicrograph was taken at 400X magnification using 400ASA Ektachrome film, courtesy of Dr. Nabil Hagag.

This book is printed on acid-free paper. ∞

Copyright © 1993 by ACADEMIC PRESS, INC.
All Rights Reserved.
No part of this publication may be reproduced or transmitted in any form or by any means, electronic or mechanical, including photocopy, recording, or any information storage and retrieval system, without permission in writing from the publisher.

Academic Press, Inc.
1250 Sixth Avenue, San Diego, California 92101-4311

United Kingdom Edition published by
Academic Press Limited
24–28 Oval Road, London NW1 7DX

Chromosome Microdissection And Cloning: A Practical Guide
Nabil Hagag and Michael V. Viola

ISBN: 0-12-313320-3

PRINTED IN THE UNITED STATES OF AMERICA
93 94 95 96 97 98   EB   9 8 7 6 5 4 3 2 1

Contents

Preface ix

Chapter 1 Introduction to Chromosome Microdissection

Chromosome Organization 1
Cloning DNA from Chromosome Fragments 14
References 16

Chapter 2 Preparation of Chromosomes for Microdissection

Introduction 25
Critical Aspects of Chromosome Preparation 26
  Enrichment of Metaphase Spreads 26
  Hypotonic Treatment 27
  Chromosome Fixing and Spreading 28
  Aging, Storing, and Staining of Metaphase Spreads 29
Reagents 30
Equipment 31
Protocols 32
  Protocol 1. Preparation of Chromosomes from Peripheral Blood T Lymphocytes (Whole Blood Microculture Method) 32
  Protocol 2. Preparation of Chromosomes from Monolayer Tissue Culture Cell Lines 34
  Protocol 3. Preparation of Chromosomes from Monolayer Cells Grown on Coverslips 35

Protocol 4. Preparation of Chromosomes from Dipteran Salivary Glands 36

Protocol 5. Solid Staining and GTG Banding of Metaphase Chromosomes 37

References 37

Chapter 3 Methods of Chromosome Microdissection

Introduction 41

Methods 42

Video Microscope Method 42

Oil Chamber Method 65

Laser Microdissection Method 67

Summary of Chromosome Microdissection and Collection for DNA Cloning 70

Reagents and Equipment 70

Protocol 70

References 73

Chapter 4 Molecular Cloning of Microdissected Chromosomal DNA

Introduction 77

Cloning DNA from Microdissected Chromosomal DNA Fragments 78

Method 1. Direct Cloning of DNA from Microdissected Chromosomal Fragments 79

Protocol 1.1. Direct Cloning into λ Phage 81

Method 2. Ligation of Microdissected Chromosomal DNA with Plasmid Vector or Linker–Adaptor and PCR Amplification 85

Protocol 2.1. Ligation of Microdissected DNA with Plasmid Vector, PCR Amplification, and Cloning   88

Protocol 2.2. Ligation of Microdissected DNA with Linker–Adaptor, PCR Amplification, and Cloning   91

Method 3. PCR Amplification of Microdissected Chromosomal DNA Fragments Followed by Probing a Complete Recombinant Library   95

Protocol 3.1. Preparation of Chromosomal DNA for Amplification   98

Protocol 3.2. PCR Amplification of Microdissected Chromosomal DNA Using "Universal" Primers   99

Protocol 3.3. PCR Amplification Using Human *Alu* Sequence-Based Primers   101

Analysis of Recombinant Clones Derived from Microdissected Chromosomal DNA   104

Determination of DNA Insert Size Range   104

Determination of the Percentage of Recombinant Clones Containing Repeat and Unique Sequences   105

Protocol 4.1. Assay for Repeat Sequences   105

Protocol 4.2. Assay for Unique Sequences   106

Calculation of the Percentage of Total Microdissected DNA Cloned   107

Determination of Potential Structural Gene Sequences   109

Localization of Recombinant Clones Using *in Situ* Hybridization   109

Protocol 5.1. DNA Probe Labeling for Fluorescence *in Situ* Hybridization   112

Protocol 5.2. Fluorescent *in Situ* Hybridization to Metaphase Chromosome Spreads   114

References   123

Chapter 5 Applications of Chromosome Microdissection

Direct Analysis of the PCR Product of Microdissected Chromosome Fragments   131

    Gene Mapping   131

    Mapping Sites of Chromosome Rearrangement and Deletions   132

    Determination of Coupling Phase   133

Recombinant DNA Libraries Generated from Microdissected Chromosome Fragments   133

    Genetic Analysis of Specialized Chromosome Structures   134

    Applications in Genomic Sequencing Projects   134

    Characterization of Disease-Related Genetic Loci   135

    Study of Chromosome Abnormalities in Cancer Cells   136

Gene Transfer Using Chromosome Fragments   136

References   139

Appendix: List of Suppliers and Addresses   143
Glossary   147
Index   155

Preface

It has been over ten years since the first report of cloning of DNA from a microdissected chromosome band. Despite the obvious usefulness of directly analyzing DNA from specific chromosome regions, there has been a paucity of publications describing experiments using chromosome microdissection. In all likelihood, such experiments have been perceived as being technically difficult and expensive. The aim of this manual is to present microdissection methodology in a straightforward manner, describing "in-house secrets" that are usually omitted from published works. We hope this manual will be particularly helpful to investigators setting up microdissection systems *de novo*.

Chromosome Microdissection and Cloning: A Practical Guide presents an overview of the current procedures and briefly reviews a few areas of research in which these techniques are being applied. The last chapter on "applications" is speculative in some respects, but with the appreciation that this methodology is still evolving. The content of this manual is divided among five chapters. Chapter 1 is a general discussion of the structure and organization of chromosomes. Chapter 2 presents methods for preparing and preserving chromosomal DNA in a manner which is useful for cloning and direct analysis. Microdissection of metaphase chromosomes and isolation of fragments can be accomplished in one of the three ways described in Chapter 3, namely, (1) by microdissection using an upright microscope and glass capillaries in an oil chamber, (2) by laser microbeam, and (3) with the use of an inverted microscope equipped with a video camera and high magnification/high resolution lenses. A step-by-step guide to these techniques and solutions for common problems are given following each method. Protocols for cloning and identifying genetic sequences from defined chromosome regions, particularly using the polymerase chain reaction, are discussed in Chapter 4. Applications of chromosome microdissection are discussed in Chapter 5. This chapter discusses current interest in using chromosome microdissection for generating "sequence tagged sites" to be used in large DNA sequencing projects, as well as using chromosome microdissection to expedite the cloning of disease-specific genes. The use of chromosome fragments for gene transfer into either tissue culture cells or fertilized ova is an area of biological interest which is still in its infancy.

We would like to thank Mary Fenstermacher for typing the manuscript and Tom Shenk and Carl Pinkert for their critiques. Our gratitude is expressed to the staff of Academic Press for their help and cooperation in the completion of this book.

Nabil G. Hagag
Michael V. Viola

Chapter 1

Introduction to Chromosome Microdissection

Chromosome microdissection is a specialized aspect of cell microsurgery. In contrast to cell microsurgical experiments, which usually are designed to measure physiological effects of cell wounding, chromosome microdissection removes and utilizes the dissected cell component. Technical advances in DNA cloning, in *in vitro* DNA amplification using the polymerase chain reaction (PCR), and in gene transfer techniques now enable biochemical and biological experiments to be performed using the small amount of DNA obtained from the dissected chromosome fragments.

In this introductory chapter, we review pertinent aspects of chromosome structure as a background for microdissection experiments. The review is followed by a brief summary of studies performed prior to 1993 that involved chromosome microdissection.

Chromosome Organization

The manner in which a linear uninterrupted double strand of DNA and its associated proteins are organized spatially in a chromosome is known in considerable detail. The primary determinants of chromosome structure include size, that is, approximately 5 cm linear DNA must be folded into a compact structure that will fit into a spherical nucleus with a volume of approximately 60 μm^3. Also, chromosomes must be of appropriate size and shape to be able to occupy specific spatial domains within the nucleus. Finally, the higher order structure of chromatin in chromosomes must allow DNA to serve as a template during DNA replication and transcription. Thus, specific DNA sequences must be accessible to a variety of polymerases, modifying enzymes, activators, and repressors. In addition, transition must occur relatively rapidly from an open, transcriptionally active form of interphase euchromatin to a more contracted chromatin, as found in transcriptionally inactive metaphase chromatin. The reverse of this process must occur when cells progress to interphase. Also, a functional and structural partition of genetic information is likely to exist that allows certain DNA regions or domains to function independently and be isolated from neighboring domains. Several models have been proposed that partially satisfy these functional constraints. These models are referred to in the subsequent discussion (1–8).

Chromosome organization usually is described by the following structures, which represent successively higher orders of DNA folding: naked double-stranded DNA, DNA coiled around core nucleosomes, the 30-nm chromatin fiber, the 250-nm chromatin fiber containing DNA loops, helical coiling of the 250-nm chromatin fiber, chromosome bands, and, finally, chromosome regions (see Figure 1.1).

Nucleosomes

DNA in chromatin is associated strongly with histone proteins. In this interaction, linear double-stranded DNA is coiled (in a curved path) around a histone core octamer (two copies each of H2A, H2B, H3, and H4). The DNA undergoes two left-handed superhelical turns around the core histone octamer. Adjacent nucleosomes are connected by "linker" DNA (9, 10). A 146-base pair (bp) nuclease-resistant segment of DNA is associated with the core particle. Linker DNA contains from 8 to 114 bp in different eukaryotic species. One mole-

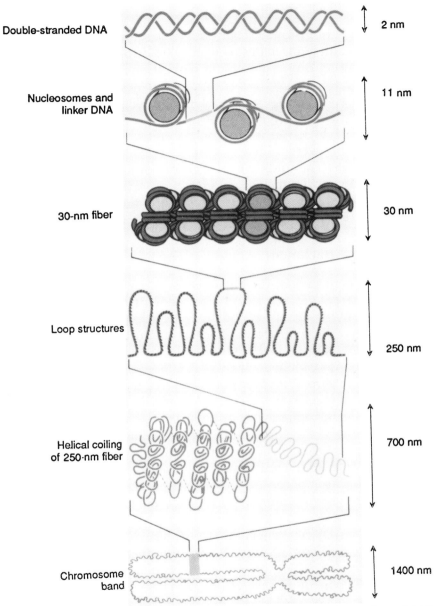

Figure 1.1 Highly schematic illustration of successively higher orders of folding of DNA in chromosomes. Adapted with permission from Alberts *et al.* (74).

cule of histone H1 appears to seal the DNA to the histone octamer. Multiple mechanisms of histone displacement from the nucleosome during transcription and replication are likely (11, 12).

30-nm Fiber

Early electron microscopic studies revealed that nucleosomes on a string were folded in a manner that generated a 30-nm chromatin fiber. The folding is in the form of a simple solenoid structure containing six nucleosomes per turn (approximately 1.2 kbp DNA) (1). Within this superhelical structure, the flat faces of the nucleosomes are tilted relative to the long axis of the 30-nm fiber.

DNA Loops: Attachment to and Coiling of a Scaffold

The details of the higher order of chromosome organization we describe as loop formation are still far from clear. In both the interphase and the metaphase nucleus, the 30-nm fiber is thought to be folded in a loop structure. The chromatin loops contain between 5 and 100 kbp of DNA. One model predicts radial arrays of ten 30-kb loops yielding an array of ~300 kb that is ~250 nm in width (7, 8). The chromatin loops are anchored to a nonhistone protein structure that is referred to as the nuclear matrix or the chromosome scaffold. The metaphase chromosome scaffold is positioned along the central axis of the chromatid. Topoisomerase II is a major component of the scaffold structure and physically extends the entire length of the chromatid. Thus, topoisomerase II appears to function catalytically during replication and transcription, but also as a "loop fastener" that secures chromatin loops to the scaffold [reviewed in Refs. (13) and (14)].

DNA segments that interact with the scaffold (scaffold-associated regions, SAR) or with the nuclear matrix (matrix-associated regions, MAR) are 0.6–1.0 kbp in length and are presumed to form the base of the loop structure. These segments contain multiple topoisomerase II cleavage consensus sequences. Matrix-associated regions frequently reside near enhancer-like sequences and can augment transcriptional activity (15, 16). When a reporter gene is flanked by MARs, the gene exhibits position-independent expression; that is, the MARs insulate the reporter gene from the effects of contiguous DNA segments (17). Thus, MARs may separate and insulate independent transcription units, possibly contained in a single loop structure.

Several lines of evidence suggest that the DNA loop may represent a distinct replication unit. The average size of DNA loops is similar to the estimated size of replicons (a region served by a single origin of replication) (18, 19). Also, origins of replication and newly replicated DNA remain attached to the nuclear matrix throughout the cell cycle (20, 21).

The metaphase scaffold can be visualized microscopically using treatments that partially deplete chromosomes of histones. Under these conditions, the scaffold has been observed to undergo a helical coiling resulting in a further ninefold compaction of the chromosome in metaphase (5, 6). This additional structural feature accounts for the zigzag appearance of chromatids that was described in earlier literature (22) and probably represented partially uncoiled structures.

During interphase, the extended, uncoiled, chromosome scaffold structure yields chromatids approximately 250 nm in width, whereas the condensed, coiled scaffold in metaphase yields chromatids 700 nm in width. After metaphase, selective decondensation (uncoiling) of euchromatic chromosome regions occurs. The entire process of chromosome condensation and decondensation involves complex interactions of nonhistone proteins with nuclear matrix structures, as well as energy-dependent phosphorylation, methylation, and acetylation reactions of histone and nonhistone chromosomal proteins (23–25).

Chromosome Bands

Specific mammalian chromosomes can be identified by the pattern of transverse bands produced by a fluorescent dye or by Giemsa stain. Chromosome bands have become topographical landmarks used to map genes, inherited traits, and chromosome structural abnormalities, and as reference points for large-scale genomic sequencing programs. Chromosome bands are known now to represent not only a structural but also a functional compartmentalization of the genome, as is summarized in the subsequent paragraphs. The significance of chromosome bands in genome organization and their evolution in plants and animals have been the subject of several reviews (7, 26–28).

A model of the structure of a human chromosome band has been described with respect to higher orders of chromatin organization and is shown in Figure 1.2 (7, 8). This model predicts that a dark G band in mid-prophase could represent from one to several stacked 300-kbp radial arrays.

The number of visible bands (e.g., Giemsa dark bands) depends on the exact stage in the cell cycle at which the chromosomes are examined. Chromosomes in mid-prophase will yield approximately 2000 bands in a human karyotype (29). Later, in early metaphase, when chromosomes have condensed, only 450–800 G bands will be visualized (30). An average dark G band in prophase contains approximately 1.5 megabase pairs (Mbp) of DNA, whereas an average size G band in metaphase contains 10–30 Mbp DNA. Therefore, condensed metaphase bands are more heterogeneous (i.e., composed of multiple band types) than those found in mid-prophase. Each metaphase G band contains ~15–50 fg DNA. Depending on the application (e.g., for cloning chromosomal DNA), 20–100 bands usually are dissected in chromosome microdissection experiments.

The first chromosome bands (Q bands) detected in humans were obtained by staining chromosomes with quinacrine mustard (31). Quinacrine mustard and its derivatives have a specificity for chromatin regions that are rich in deoxyadenine (dA) and deoxythymidine (dT), and produce a different banding pattern than fluorochromes with affinity for deoxycytidine (dC) and deoxyguanidine (dG) residues (e.g., chromomycin A_3, mithramycin). Treatment of fixed chromosomes with trypsin followed by staining with Giemsa (GTG banding, as described in Chapter 2) yields dark (G) and light (reverse or R) bands. R bands usually correlate with bright bands detected with quinicrine (Q bands). A reference human karyotype showing G and R bands at the 550-band level is shown in Figure 1.3. Giemsa staining of human chromosomes after heat treatment results in preferential

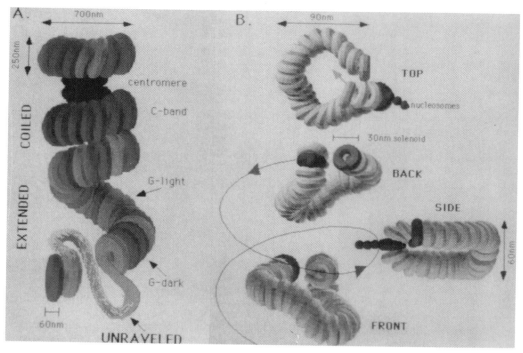

Figure 1.2 (A) Model of higher order structure of interphase chromosome fiber. The 250-nm chromatin fiber is highly coiled through the heterochromatic centromere and paracentromeric region. In this model, the 250-nm fiber is composed of stacks of 300-kb radial arrays of loop structures. In the lower part of the figure, a G-light radial array is unraveled, as might occur during transcription. (B) Proposed folding of 30-nm solenoid fiber into a "radial array," which is the subunit of the 250-nm fiber. Detailed discussion is found in Refs. (7) and (8). [Reprinted with permission from Manuelides (8).]

staining of centromeric and paracentromeric regions, as well as regions of chromosomes 1, 9, 16 and the telomere of the Y chromosome; these patterns are referred to as C bands. C bands are regions of constitutive heterochromatin and remain condensed throughout the cell cycle, except for a short period of time during DNA replication. The major staining techniques likely to generate bands to be used

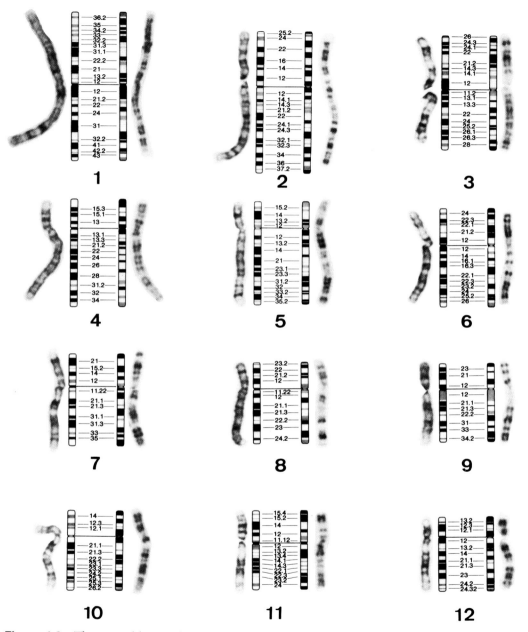

Figure 1.3 The normal human karyotype demonstrating G and R banding at the 550-band stage. [Reprinted with permission from Francke (75) and Camargo and Cervinka (76).]

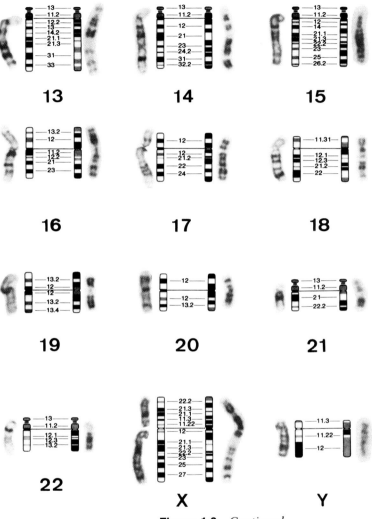

Figure 1.3 *Continued*

in microdissection experiments are summarized in several excellent cytogenetic manuals (32, 33) and in Chapter 2.

The structural compartmentalization of chromosomes into transverse bands, as seen with Giemsa and fluorescent dyes, can also be seen by treating chromosome spreads with the restriction endonuclease *Hae*III, as well as with *in situ* hybridization using repeat-sequence DNA probes. Long interspersed repeats (LINES) are found predominantly in G bands, whereas short interspersed repeats (SINES or *Alu* sequences) are found in R bands (34, 35). The differing base composition of these two repeat DNA groups accounts for the higher GC content of R bands in contrast to the higher AT content of G bands. The distribution of LINES and SINES in bands may be a major determinant of the banding pattern produced by fluorescent dyes that have base specificity when they bind. C bands do not contain SINES or LINES as found in R and G bands, respectively, but are made of simple-sequence, highly repetitive, satellite DNA sequences (36).

Not surprisingly, the structural organization of chromosome bands has profound implications for the organization and regulation of gene expression. Most (if not all) housekeeping genes and the majority of tissue-specific genes are found in R bands. In this regard, clustering of nonmethylated CpG dinucleotides (HTF islands), which usually are found 5′ to "housekeeping" genes, appears restricted to R bands (24). Further, expressed genes replicate early in S phase whereas nonexpressed genes replicate late. Using bromodeoxyuridine (BrdU) labeling of chromosomes (which generates replication bands), DNA in R bands has been shown to complete replication before DNA in G bands begins replication. The structural and functional inverse relationships between different chromosome bands are summarized in Table 1.1.

Polytene Chromosome Bands

The replication of DNA in the absence of subsequent cell division results in multiple copies of normal diploid DNA (polyploidy). In certain insect secretory cells, the polyploid DNA is in the form of

Table 1.1 Structural and Functional Characteristics of Chromosome Bands

G Bands	R Bands
Quinacrine dull	Quinacrine bright
Heterochromatic	Euchromatic
A + T rich	G + C rich
Short interspersed repetitive DNA (SINES)	Long interspersed repetitive DNA (LINES)
Alu repeat sequences	L1 repeat sequences
Few CpG islands	Contains CpG islands
Few tissue-specific genes	Most tissue-specific and housekeeping genes
Late replicating	Early replicating

multiple copies of entire chromatids that are longitudinally aligned side by side (polytene chromosomes; 37, 38). In the dipteran salivary gland, the chromosome arm exists as 2^{10} or 1024 copies. Horizontal bands are visible on polytene chromosomes without prior staining and represent regions of chromatin condensation. Approximately 5000 bands and an equal number of interband regions can be seen. Multiple genes appear to reside in a single band. Polytene chromosome bands are ideal for microdissection. Since the chromosomes are redundant, fewer fragments need to be collected. An average size band (~0.1–0.2 μm) contains ~30 kbp and yields 0.01–0.1 pg DNA.

Chromosome Regions

The primary constriction in mammalian chromosomes is called the centromere; its location is useful in the preliminary grouping of human chromosomes. Centromeres located centrally yield two chromosome arms of equal length (metacentric); the centromere position may yield two unequal chromosome arms (submetacentric) or be near (acrocentric) or at the end (telocentric) of the chromosome. The structure of centromeres and chromosome ends (telomeres) has been the subject of considerable investigation (39–43) and is discussed in the following sections. The centromere and telomere are regions that are very amenable to microdissection.

Centromeres

The centromere and its specialized structure, the kinetochore, play an important role in sister chromatid separation through attachment of the kinetochore to microtubules associated with the spindle apparatus. The centromere is also the region of sister chromatid pairing and may be the site of the molecular "motor" responsible for chromosome segregation during anaphase.

The primary constriction results from a lack of coiling of the 250-nm chromatin fiber in that region (Figure 1.4). The higher order structure of the centromere has been revealed using drugs (e.g., 33258 Hoechst, 5-azacytidine) that have a specific affinity for the centromeric region. The kinetochore may be divided into a number of domains based on its electron microscopic appearance (Figure 1.4). Specific antibodies isolated from patients with autoimmune disorders have defined a number of centromeric proteins, as well as kinetochore-associated proteins [reviewed in Ref. (39)].

The centromere is composed of constitutively condensed chromatin; satellite DNA plays an important role in its structure. The satellite sequences also present a barrier to recombination events. Minor satellite DNA is localized to discrete regions of the mouse centromere, suggesting a functional organization of centromere repeat sequences (44). The predominant satellite family in the human centromere is alpha satellite DNA, composed of multiple copies of a highly diverging basic repeat unit of approximately 171 bp [reviewed in Ref. (36)]. Tandem repeats of alpha satellite DNA are further organized into macro repeat units ranging from 0.5 to 10 Mbp in length. Most human chromosomes can be distinguished from one another by their divergent alpha satellite sequences (for which probes are now available commercially) and by the characteristic number of tandem repeats. Other families of human satellite DNA, for example, classical

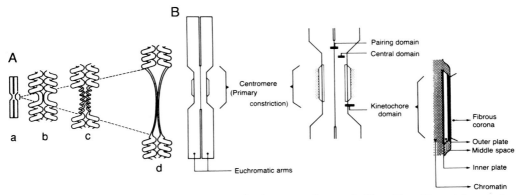

Figure 1.4 Structure of a centromere. (A) Path of 250-nm chromatin fiber. Uncoiling of chromatin fiber at centromere when cells are grown in 33258 Hoechst (a–d). (B) Diagram of structural domains of a centromere. [Reprinted with permission from Rattner (39).]

satellite DNA (45) and beta satellite DNA (46), appear to be localized to the pericentromeric region.

Telomeres

For many years, the ends of chromosomes have been known to contain specialized structures that are essential for maintaining chromosome integrity. Chromosomes with broken tips either degrade or fuse to the ends of other broken chromosomes. The chromosome ends, or telomeres, are likely to be important in spatial positioning of the chromosome in the nucleus as well.

The primary structure of the telomere is a conserved G/C-rich repeat unit of the general formula $D(T/A_{1-4} dG_{1-8})$. The characteristic consensus sequence found in *Tetrahymena* (dTTGGGG) is quite similar to that found in humans (dTTAGGG) (43). The terminal repeat number varies considerably in different species, from a few hundred base pairs in ciliates to 10 kbp in humans, and up to 150 kbp in mice. A significant common feature of all telomeres is an overhanging

G-rich DNA strand that extends 12–16 bp beyond the C-rich strand. Single-stranded G-rich oligonucleotides can form unusual folded structures in solutions, for example, hairpin duplexes (by G · G non-Watson–Crick pairing) and four stranded structures. Whether these structures exist *in vivo* is unclear (47, 48). However, since telomeres do not have covalently bound terminal proteins or Watson–Crick paired terminal hairpins, as found in some linear viral genomes, the 3′ G-rich overhang is likely to assume a folded configuration that protects the terminus.

A specialized reverse transcriptase, telomerase, adds telomeric repeats to the 3′ chromosome end. The enzyme has an associated RNA molecule that serves as template for *de novo* synthesis of the repeat on the G-rich strand (49).

Cloning DNA from Chromosome Fragments

The first experiments that successfully generated recombinant genomic DNA clones from microdissected chromosome fragments were performed in the early 1980s by Jan-Erik Edström and his colleagues at the European Molecular Biology Laboratory in Heidelberg (50). The methodology was an outgrowth of the microchemistry work that Edström had been doing since the 1960s (51). *Drosophila* polytene chromosomes squashed onto coverslips were inverted over an oil chamber and chromosomes were dissected using bent, microforged, glass-cutting micropipettes. Fragments were pooled into a nanoliter-sized drop of aqueous solution in the oil. The subsequent biochemical steps necessary for cloning (phenol/chloroform extraction, restriction enzyme digestion, ligation to cleaved vector) also were performed in the aqueous drop. The small quantity of recombinant DNA then was added to a packaging mixture and recombinant phage were replicated in *Escherichia coli*. Using this technique (microdissection and microcloning), genomic clones were obtained from *Drosophila melanogaster* polytene chromosomes (50–53), *Drosophila hydei* lampbrush DNA loops (54), and mouse (55–59) and human (60–63) metaphase chromosomes. In these studies, the generation of multiple clones from specific genetic regions expedited the identification and isolation of a number of important genes, for example, the *t* complex in mice (55) and the *Krüppel* gene in *Drosophila* (52). However, microcloning generates a library of recombinant clones with small DNA inserts (usually <1 kbp), which represents only a small fraction (<5%) of the total DNA sequences in the microdissection region. These limitations are a result of the depurination of DNA that occurs during chromosome fixation and of the inefficiency of cloning small quantities of genomic DNA. An additional problem of considerable importance in some studies is the generation of recombinant clones that are not derived from the microdissected chromosome fragments and are most likely acquired from contaminating DNA (e.g., bacterial) on slides and glassware.

Several technical advances have improved both the facility in performing precise chromosome dissection and the efficiency of cloning dissected DNA. First, procedures for preparing metaphase chromosomes have been developed that minimize DNA degradation (61, 62; reviewed in Chapter 2). When care is taken to avoid prolonged acid fixation, microcloning yields recombinant clones with larger DNA inserts. Second, as an alternative to cutting chromosomes in oil chambers using an upright microscope, chromosome spreads may be cut in air (or under a drop of oil) using an inverted microscope and straight glass-cutting pipettes (64; see Chapter 3). Further, higher magnification and resolution of minor chromosome bands can be obtained using video microscopy. Finally, a series of cloning procedures has been devised that uses PCR and attempts to improve the poor efficiency of microcloning. Microdissected chromosome DNA has been ligated enzymatically either to oligonucleotide linker primers (65) or to cleaved plasmids (66). The subsequent PCR reaction utilizes primer sequences present either in the linker primer or, in the latter instance, in the vector. DNA from several human chromosome bands of considerable clinical interest has been cloned in this way (67–71). This method generates a large number of recombinant clones (usually >20,000) with very small inserts. Use of restriction enzymes that cut DNA frequently and degradation of DNA during chromosome fixation probably account for the small insert size.

Although amplification of specific sequences appears to have no bias, smaller DNA fragments may be copied preferentially. Despite the large number of clones, the small insert size and redundancy of clones results in cloning of only a small fraction of the microdissected DNA. An alternative method likely to yield a more complete chromosomal library employs PCR amplification of microdissected chromosomal DNA using either "universal" random primers (72) or *Alu* sequence primers (73; see Chapter 4). The amplification product is used to probe a complete genomic or chromosomal library. The selected library contains clones with large inserts. If the PCR reaction is symmetrical and unbiased, a more complete library can be obtained with this method than with other methods.

References

1. Finch, J., and Klug, A. (1976). Solenoid model for suprastructure in chromatin. *Proc. Natl. Acad. Sci. USA* **73,** 1897–1901.

2. Adolph, K. W. (1980). Organization of chromosomes in mitotic HeLa cells. *Exp. Cell Res.* **125,** 95–103.

3. Belmont, A. S., Braunfield, M. B., Sedat, J. W., and Agard, D. A. (1989). Large scale structural domains within mitotic and interphase chromosomes *in vivo* and *in vitro. Chromosoma* **98,** 129–143.

4. Laemmli, U. K., Cheng, S. M., Adolph, K. W., Paulson, J. R., Brown, J. A., and Baumback, W. R. (1978). Metaphase chromosome structure: The role of nonhistone proteins. *Cold Spring Harbor Symp. Quant. Biol.* **42,** 109–118.

5. Rattner, J. B., and Lin, C. C. (1985). Radial loops and helical coils coexist in metaphase chromosomes. *Cell* **41,** 291–296.

6. Boy de la Tour, E., and Laemmli, U. K. (1988). The metaphase scaffold is helically folded: Sister chromatids have predominantly opposite helical handedness. *Cell* **55,** 937–944.

7. Manuelides, L., and Chen, T. L. (1990). A unified model of eukaryotic chromosomes. *Cytometry* **11,** 8–25.

8. Manuelides, L. (1990). A view of interphase chromosomes. *Science* **250,** 1535–1540.

9. Richmond, T. J., Finch, J. T., Reishton, B., Rhodes, D., and Klug, A. (1984). Structure of the nucleosome core particle at 7 Å resolution. *Nature* **311,** 532–537.

10. McGhee, J. O., and Felsenfeld, G. (1980). Nucleosome structure. *Annu. Rev. Biochem.* **59,** 1115–1156.

11. Thoma, F. (1991). Structural changes in nucleosomes during transcription: Strip, split or flip. *Trends Genet.* **7,** 175–177.

12. Kornberg, R. D., and Lorch, Y. (1991). Irresistible force meets unmovable object: Transcription and the nucleosome. *Cell* **67,** 833–836.

13. Gasser, S. M., and Laemmli, U. K. (1987). A glimpse at chromosomal order. *Trends Genet.* **3,** 16–22.

14. Nelson, W., Pienta, K. J., Barrock, E. R., and Coffey, D. S. (1986). The role of the nuclear matrix in the organization and function of DNA. *Annu. Rev. Biophys. Chem.* **15,** 457–475.

15. Cockerill, P. N., and Garrard, W. T. (1986). Chromosome loop anchorage of the kappa immunoglobulin occurs next to the enhancer in a region containing topoisomerase II sites. *Cell* **44,** 273–282.

16. Stief, A., Winter, D. M., Strotling, W. H., and Sippel, A. E. (1989). A nuclear DNA attachment element mediates elevated and position independent gene activity. *Nature* **341,** 343–345.

17. Kellum, R., and Schedl, P. (1991). A position-effect assay for bounderies of higher order chromosome domains. *Cell* **69,** 941–950.

18. Buongiorno-Nardelli, M., Micheli, G., Carri, M. T., and Maulley, M. (1982). A relationship between replicon size and supercoiled loop domains in the eukaryotic genome. *Nature* **298,** 100–102.

19. Sundin, O., and Varshausky, A. (1980). Terminal stages of SV40 DNA replication proceed via multiply intertwined catenated dimers. *Cell* **21,** 103–114.

20. Van der Velden, H. M. W., van Willigen, G., Wetzels, R. H. W., and Wanka, F. (1984). Attachment of origins of replication to the nuclear matrix and chromosomal scaffold. *FEBS Lett.* **161,** 13–16.

21. McCready, S. J., Goodwin, J., Mason, D. W., Brazel, I. A., and Cook, P. R. (1980). DNA is replicated at the nuclear cage. *J. Cell Sci.* **46,** 365–386.

22. Onnuki, Y. (1968). Structure of chromosomes. I. Morphological studies of the spiral structure of human somatic chromosomes. *Chromosoma* **25,** 402–408.

23. Lamb, N., Fernandez, A., Watrin, A., Labbe, J.-C., and Cavadore, J.-C. (1990). Microinjection of p34^{cdc2} kinase induces marked changes in cell shape, cytoskeletal organization, and chromatin structure in mammalian fibroblasts. *Cell* **60,** 151–165.

24. Hebbes, T., Thorne, A., and Ciane-Robinson, C. (1988). A direct link between core histone acetylation and transcriptionally active chromatin. *EMBO J.* **7,** 1395–1402.

25. Bird, A. P. (1986). CpG-rich islands and the function of DNA methylation. *Nature* **321,** 209–213.

26. Bickmore, W. A., and Sumner, A. T. (1989). Mammalian chromosome banding—An expression of genome organization. *Trends Genet.* **5,** 144–148.

27. Sumner, A. T. (1990). "Chromosome Banding." Unwin Hyman, New York.

28. Holmquist, G. P. (1989). Evolution of chromosome bands: Molecular ecology of non-coding DNA. *J. Mol. Evol.* **28,** 469–486.

29. Yunis, J. (1981). Mid-prophase human chromosomes. The attainment of 2000 bands. *Hum. Genet.* **56,** 293–298.

30. Harnden, D. G., and Klinger, H. P. (eds.) (1985). "An International System for Human Cytogenetic Nomenclature." Karger, Basel.

31. Caspersson, T., Farber, S., Kudynowski, G. E., Modest, E. J., Simonsson, E., Wagh, U., and Zech, L. (1986). Chemical differentiation along metaphase chromosomes. *Exp. Cell Res.* **49,** 219–222.

32. Rooney, D. E., and Czepulkowski, B. H. (eds.) (1986). "Human Cytogenetics: A Practical Approach." IRL Press, Oxford.

33. Macgregor, H. C., and Vailey, J. C. (eds.) (1988). "Working with Animal Chromosomes." Wiley, New York.

34. Manuelides, L., and Ward, D. C. (1984). Chromosomal and nuclear distribution of the *Hind*III 1.9-kb human DNA repeat segment. *Chromosoma* **91,** 28–38.

35. Korenberg, J. R., and Rykowski, M. C. (1988). Human genome organization: ALU, LINES, and the molecular structure of metaphase chromosome bands. *Cell* **53,** 391–400.

36. Willard, H. (1990). Centromeres of mammalian chromosomes. *Trends Genet.* **6,** 410–0416.

37. Hill, R. J., and Rudkin, G. (1987). Polytene chromosomes: The status of the band–interband question. *BioEssays* **7,** 35–40.

38. Beerman, W. (1972). Chromosomes and genes. *In* "Developmental Studies on Giant Chromosomes" (W. Beerman, ed.). Springer-Verlag, New York.

39. Rattner, J. B. (1991). The structure of the mammalian centromere. *BioEssays* **13,** 51–56.

40. Pluta, A. F., Cooke, C. D., and Earnshaw, W. C. (1990). Structure of the human centromere at metaphase. *Trends Biol. Sci.* **15,** 181–185.

41. Moyzes, R. (1991). The human telomere. *Sci. Am.* **245,** 48–57.

42. Zakian, V. A. (1989). Structure and function of telomeres. *Annu. Rev. Genet.* **23,** 579–604.

43. Blackburn, E. (1991). Structure and function of telomeres. *Nature* **350,** 569–573.

44. Radic, M., Lundgren, K., and Hamkalo, B. (1987). Curvature of mouse satellite DNA and condensation of heterochromatin. *Cell* **50,** 1101–1108.

45. Moyzis, R. K., Albright, K. L., Bartholdi, M. F., Cram, L. S., Devin, L. I., Hildernbrand, C. E., Joste, W. E., Longmire, J. E., Meyne, J., and Schwartzcher-Robinson, T. (1987). Human chromosome-specific repetitive DNA sequences: Novel markers for genetic analysis. *Chromosoma* **95,** 375–386.

46. Wayne, J. S., and Willard, H. (1989). Human beta satellite DNA: Genomic organization and sequence definition of a class of highly repetitive tandem DNA. *Proc. Natl. Acad. Sci. USA* **86,** 6250–6254.

47. Williamson, J., Raghuraman, M. K., and Cech, T. (1989). Monovalent cation-induced structure of telomeric DNA: The G-quartet model. *Cell* **59,** 871–880.

48. Henderson, E., Harden, C., Walk, S., Tinoco, L., and Blackburn, E. (1987). Telomeric DNA oligonucleotides form novel intramolecular structures containing guanine–guanine base pairs. *Cell* **51,** 899–9098.

49. Greider, C. W., and Blackburn, E. H. (1985). Identification of a specific telomere terminal transferase activity in *Tetrahymena* extracts. *Cell* **43,** 405–413.

50. Scalenghe, F., Turco, E., Edström, J. E., Pirotta, V., and Melli, M. (1981). Microdissection and cloning of DNA from a specific region of *Drosophila melanogaster* polytene chromosomes. *Chromosoma* 82, 205–216.

51. Edström, J.-E. (1964). Microextraction in microelectroporesis for determination and analysis of nucleic acids in isolated cellular units. *In* "Methods in Cell Physiology" (D. M. Prescot, ed.), pp. 417–447. Academic Press, New York.

52. Preiss, A., Rosenberg, U. B., Kienlin, A., Seifert, E., and Jackle, H. (1985). Molecular genetics of *Krüppel*, a gene required for segmentation of the *Drosophila* embryo. *Nature* 313, 27–37.

53. Johnson, D. H., Edström, J.-E., Burnett, J. B., and Friedman, T. B. (1987). Cloning of a *Drosophila melanogaster* adenine phosphoribosyltransferase structural gene and deduced amino acid sequence of the enzyme. *Gene* 59, 77–86.

54. Hennig, W., Huijser, P., Vogt, P., Jackle, H., and Edström, J.-E. (1983). Molecular cloning of microdissected lampbrush loop DNA sequencing of *Drosophila hydei*. *EMBO J.* 2, 1741–1746.

55. Röhme, D., Fox, H., Hermann, B., Frischauf, A.-M., Edström, J.-E., Maires, P., Silver, L. M., and Lehrach, H. (1984). Molecular cloning of the mouse t complex derived from microdissected metaphase chromosomes. *Cell* 36, 783–788.

56. Fischer, E. M. C., Cavanna, J. S., and Brown, S. D. M. (1985). Microdissection and microcloning of the mouse X chromosome. *Proc. Natl. Acad. Sci. USA* 82, 5846–5849.

57. Brockdorff, N., Fischer, E. M. C., Cavanna, J. S., Lyon, M. F., and Brown, S. D. M. (1987). Construction of a detailed molecular map of the mouse X chromosome by microcloning and interspecific crosses. *EMBO J.* 6, 3291–3297.

58. Greenfield, A. J., and Brown, S. D. M. (1987). Microdissection and microcloning from the proximal region of mouse chromo-

some 7; isolation of genes genetically linked to Pudgy locus. *Genomics* **1**, 153–158.

59. Weith, A., Winking, H., Brackmann, B., Boldyieff, B., and Traut, W. (1987). Microclones from a mouse germ line HSR, detection of amplification, and complex rearrangements of DNA sequences. *EMBO J.* **6**, 1295–1300.

60. Bates, G. P., Wainwright, B. J., Williamson, N., and Brown, S. D. (1986). Microdissection of and microcloning from the short arm of human chromosome 2. *Mol. Cell. Biol.* **6**, 3826–3830.

61. Kaiser, R., Weber, J., Grzeschik, K.-H., Edström, J.-E., Driesel, A., Zengerling, S., Buchwald, M., Tsui, L. C., and Olek, K. (1987). Microdissection and microcloning of the long arm of human chromosome 7. *Mol. Biol. Rep.* **12**, 3–6.

62. Martinsson, T., Weith, A., Cziepluch, C., and Schwab, M. (1989). Chromosome 1 deletions in human neuroblastomas: Generation and fine mapping of microclones from the distal 1p region. *Genes Chrom. Cancer* **1**, 67–78.

63. Weber, J., Weith, A., Kaiser, R., Grzeschik, K.-H., and Olek, K. (1990). Microdissection and microcloning of human chromosome 7q 22-32 region. *Som. Cell Mol. Gen.* **16**, 123–128.

64. Senger, G., Lüdecke, H.-J., Horsthemke, B., and Claussen, U. (1990). Microdissection of banded human chromosomes. *Hum. Genet.* **84**, 507–511.

65. Johnson, D. (1990). Molecular cloning of DNA from specific chromosomal regions by microdissection and sequence-independent amplification of DNA. *Genomics* **6**, 243–251.

66. Lüdecke, H.-J., Senger, G., Claussen, U., and Horsthemke, B. (1990). Construction and characterization of band-specific DNA libraries. *Hum. Genet.* **84**, 512–516.

67. Lüdecke, H.-J., Senger, G., Claussen, U., and Horsthemke, B. (1989). Cloning defined regions of the human genome by microdissection of banded chromosomes and enzymatic amplification. *Nature* **338**, 348–350.

68. MacKinnon, R. N., Hurst, M. C., Bell, M. V., Watson, J. E. V., Claussen, U., Lüdecke, H. J., Senger, G., Horsthemke, B., and Davies, K. E. (1990). Microdissection of the fragile × region. *Am. J. Hum. Genet.* **47**, 181–187.

69. Buiting, K., Neuman, M., Lüdecke, H.-J., Senger, G., Claussen, U., Antich, J., Passarge, E., and Horsthemke, B. (1990). Microdissection of the Prader–Willi syndrome chromosome region and identification of potential gene sequences. *Genomics* **6**, 521–527.

70. Trautmann, U., Leuteritz, G., Senger, G., Claussen, U., and Bollhausen, W. (1991). Detection of APC region-specific signals by nonisotopic chromosomal in situ suppression (CISS) hybridization using a microdissection library as probe. *Hum. Genet.* **87**, 495–497.

71. Newsham, I., Claussen, U., Tridecke, H.-J., Mason, M., Senger, G., Horsthemke, B., and Cavenee, W. (1991). Microdissection of chromosome band 11p15.5: Characterization of probes mapping distal to the HBBC locus. *Genes Chrom. Cancer* **3**, 108–116.

72. Wesley, C., Ben, M., Kreitman, M., Hagag, N., and Eanes, W. F. (1990). Cloning regions of the *Drosophila* genome by microdissection of polytene chromosome DNA and PCR with nonspecific primer. *Nucleic Acids Res.* **18**, 599–603.

73. Guan, X.-Y., Meltzer, P., Cao, J., and Trent, J. (1992). Rapid generation of region specific genomic clones by chromosome microdissection: Isolation of DNA from a region frequently deleted in malignant melanoma. *Genomics* **14**, 680–684.

74. Alberts, B., Bray, D., Lewis, J., Roff, M., Roberts, K., and Watson, J. (eds.) (1989). "Molecular Biology of the Cell." Garland Publishing, New York.

75. Francke, U. (1981). High resolution ideogram of trypsin–Giemsa banded human chromosomes. *Cytogenet. Cell Genet.* **31,** 24–32.
76. Camargo, M., and Cervinka, J. (1982). Pattern of DNA replication of human chromosomes. *Am. J. Hum. Genet.* **34,** 757–780.

Chapter 2

Preparation of Chromosomes for Microdissection

Introduction

During the last 30 years, cytogeneticists have evolved empirically the optimum methods for high-resolution staining of chromosome bands. However, these methods must be modified when preparing chromosomes for microdissection because conventional banding techniques result in degradation of chromosomal DNA. In fact, evidence suggests that improvement in GTG banding (G bands by trypsin using Giemsa) is associated with DNA degradation (1). Several factors contribute to DNA breakdown in the preparation of chromosome spreads. DNA appears to be more unstable (i.e., susceptible to denaturation and DNAse digestion) in condensed metaphase chromatin than in interphase chromatin (2). Further, at each step in the preparation of metaphase chromosomes, for example, hyotonic swelling, fixing, and "aging," DNA degradation is facilitated (1, 3–5). Of these steps, fixation and "aging" contribute most significantly to generating chromosomal DNA that is unsuitable for biochemical and biological studies. Therefore, considering chromosomes as units of linear DNA rather than as cytogenetic structures is essential in microdissection experiments. As a result, some sacrifice in banding quality must be made.

In this chapter, we first discuss the critical aspects of chromosome preparation. Subsequently, we present detailed protocols that are most likely to be used in microdissection experiments. Reference should be made to one of several standard technical manuals used for routine chromosome preparation (6–8). Rigorous comparisons have not been made to define the optimum procedures for preserving the quality of the DNA in chromosome spreads. However, we describe methods that have been useful in our laboratory and in other laboratories involved in similar work.

Critical Aspects of Chromosome Preparation

Enrichment of Metaphase Spreads

Having as many metaphase spreads as possible on one coverslip is ideal. Only a few of the spreads are dissectable because of poor extension of some spreads, disadvantageous orientation of the chromosome of interest, or overlapping chromosomes. Short- or long-term monolayer cell cultures, as well as suspension cultures, may be synchronized in the cell cycle to capture the maximum number of cells in mitosis. Thymidine "block" and "release" is the most frequent method used and is described in the protocol section (6, 9). We avoid the use of antimetabolites (e.g., methotrexate), cytotoxic drugs (e.g., actinomycin D), or other chemicals capable of inducing DNA damage as synchronizing agents. In most instances, adequate numbers of mitotic cells can be obtained from monolayer cultures without synchronization by plating cells sparsely and harvesting mitotic cells when cells are in a log phase of growth (3–4 days later).

Mitotic cells in monolayer cultures usually become rounded and are weakly adherent to the culture dish surface. A high yield of mitotic cells (>90%) can be harvested during and after Colcemid treatment by shaking mitotic cells off the culture plate or, if cells are growing in a roller bottle, by increasing the rate of rotation of the roller bottle cultures to expel mitotic cells into the medium (10). A more tedious, but effective, method is identifying rounded mitotic cells using an inverted microscope and drawing the cells into a pipette to pool them in hypotonic solution (11).

Metaphase spreads of peripheral blood T cells may be obtained using whole blood cultures or cultures of purified lymphocytes (6). Ficoll gradient separation of leukocytes, followed by culture in medium containing phytohemagglutinin for 72 hours, yields large numbers of mitotic cells. Metaphase spreads from T lymphocyte can be obtained easily from whole blood cultures, but multiple washes in fixative are necessary to rid the cultures of red blood cells. Prolonged exposure to fixative must be avoided in this method (see protocol).

Although several effective mitotic spindle inhibitors are available, Colcemid (deacetylmethyl colchicine) is used routinely in most laboratories. Colcemid arrests cells at the M/G_2 interphase. Since chromosomes become more condensed and shorter at the end of metaphase, short incubation times with Colcemid (10 minutes–1 hour) tend to yield more extended chromosomes, which are preferable, particularly when dissecting small chromosomes or when attempting to dissect minor bands. We avoid the use of agents (e.g., ethidium bromide) that promote chromosome extension since they enhance DNA nicking.

Hypotonic Treatment

After Colcemid treatment, cells are placed in a hypotonic solution; usually 0.075 M KCl suffices. The hypotonic solution swells the nucleus and cytoplasm, breaks intrachromosomal connections, and allows a better separation of chromosomes when the cells are smashed on a coverslip. Although hypotonic treatment of cells has resulted in degradation of bulk DNA (3), this degradation has not been observed for DNA from chromosome spreads (1). Nevertheless, the shortest length of exposure to hypotonic solution that will ensure large numbers of well-spread chromosomes is preferable. This time period will vary among cell lines and should be determined experimentally. Lymphocyte cultures usually can be treated for 10 minutes at 37°C to produce adequate spreads, whereas monolayer tumor cell lines usually are treated for 30 minutes at 37°C. Prolonged hypotonic treatment results in cell lysis and dispersion of chromosomes over a large area. Less than optimum hypotonic treatment results in tightly packed, nondissectable metaphase spreads.

Chromosome Fixing and Spreading

The purpose of the fixation step is fixing chromosomes at a specific stage in the cell cycle with little distortion of morphological detail. Optimum fixation also rids the spreads of cytoplasmic debris. The most commonly used fixatives are mixtures of methanol and acetic acid, usually at a ratio of 3 : 1 or higher. Acetic acid is a small molecule with excellent penetrability that will precipitate nucleoprotein, extract considerable amounts of chromosomal proteins, and remove cytoplasmic constituents from metaphase spreads. Mixtures of acetic acid and methanol produce minimal swelling and condensation of chromosomes.

The major adverse effect of acetic acid fixation for chromosome microdissection studies is that the treatment causes depurination and nicking of DNA (4, 5, 12). Depurinated DNA is more susceptible to enzymatic digestion (13) as well as to the DNA degradation associated with chromosome "aging" (1). Therefore, a number of precautions and modifications of standard fixation protocols should be observed:

1. Acetic acid and methanol should be free of water since acid depurination is a hydration reaction. Reagents should be aliquoted into small bottles and kept well sealed.

2. All steps in chromosome fixation and spreading should be done at 0–4°C.

3. Chromosomes should be exposed to fixative for only a brief time period, which is accomplished conveniently by processing small aliquots of cells at one time. Cells are removed from the hypotonic step, sedimented by centrifugation, and resuspended in methanol:acetic acid fixative (3 : 1) at 0°C. Depending on the cell type, cells may be kept in fixative for several seconds or up to 10

minutes, resulting in rupture of nuclei when the cells are dropped onto coverslips.

Alternatively, larger cell pellets may be fixed in methanol. Cells are then dropped onto coverslips and briefly flooded with 3 : 1 methanol : acetic acid. Also, aliquots of methanol-treated cells may be sedimented, resuspended in fixative, and dropped onto coverslips (14). The "pipette" method has been used to draw cells up into a pipette filled with fixative, followed subsequently by dropping onto coverslips (11).

After brief fixation, cells are dropped from a height of approximately 2 feet onto wet, cold coverslips. Coverslips are kept at $-70°C$ and dipped into cold distilled water immediately before use. We use large coverslips (35 × 50 mm), which will accommodate 6–7 drops of fixed cells. The drops are spread across the surface of the coverslip by mouth blowing and, when almost dry, fixative is washed off by briefly placing the coverslip in distilled water. Unstained spreads are examined immediately by phase microscopy to assess if the cell concentration is optimum. The coverslip fits into a customized holder for microdissection (see Chapter 3).

Aging, Storing, and Staining of Metaphase Spreads

Once metaphase spreads are fixed and dropped onto coverslips, the chromosomes must be "aged" to obtain GTG banding. Aging is accomplished in clinical cytogenetic laboratories by keeping spread chromosomes at room temperature for 3–5 days or at 56°C overnight. The precise physicochemical basis for the aging of chromosomes is unknown, but evidently chromosomal DNA that has become depurinated by acid fixation becomes increasingly single stranded (13) and degraded (1) during the aging process. Therefore, when possible, dissecting fresh uniformly stained (solid staining) or unstained chromosomes is preferable. Staining can be avoided when dissecting telomeric or centromeric regions of chromosomes that are easily recognizable by size or morphology or when dissecting characteristic marker chromosomes. Alternatively, GTG bands can be identified on chromosome spreads stored in cold ethanol for as brief a period as 1 hour. Dissecting chromosomes within 24 hours of placing coverslips in ethanol is preferable; we have detected moderate degradation of chromosomal DNA in spreads on coverslips stored in ethanol overnight (P. Wang and M. V. Viola, unpublished data). We have found that chromosomes on coverslips that are quick-frozen in liquid nitrogen, followed by storage in air-tight boxes at $-70°C$, as is done commonly for *in situ* hybridization studies, also yield poor quality DNA (P. Wang and M. V. Viola, unpublished data). Once a coverslip is air-dried, DNA will begin "aging." Therefore, during microdissection experiments, we discard dried coverslips after 30 minutes to 1 hour and microdissect chromosomes on a new coverslip, immediately removed from ethanol and freshly stained.

Reagents

Only one commercial distributor is listed. However, for most items multiple suppliers can provide reagents of comparable quality.

Acetic acid, glacial, reagent grade (Fisher #A-38C-212)
Antibiotics, for tissue culture; penicillin–streptomycin mixture (GIBCO H43-5140)
Buffer, pH 6.8—0.01 M phosphate buffer (Na_2HPO_4, KH_2PO_4), pH 6.8; add 1 buffer tablet per liter distilled water (Biomedical Specialties #339920)
Chromosome 4A medium with phytohemagglutinin (GIBCO #123-1674AP); reconstitute with diluent as recommended by supplier
Colcemid, stock solution—10 μg/ml (GIBCO #126-5211AD) in Hank's buffered salt solution; store at $-20°C$
Distilled water, ice-cold
Dulbecco's phosphate-buffered saline solution (PBS) (GIBCO #310-428-4280AG)
Ethanol, absolute
Ficoll–Paque (Pharmacia LKB 4A 17-0840-02)

Giemsa stain—mix 1 ml Gurr Giemsa R66 with 45 ml buffer (pH 6.8) (Biomedical Specialties)

Growth medium:
  100 ml RPMI 1640 (GIBCO #R15-0576)
  23 ml fetal bovine serum, heat inactivated (GIBCO #230-6010AG)
  1.3 ml 100× penicillin–streptomycin solution (10,000 units penicillin; 10,000 μg streptomycin per ml) (GIBCO #600-5075AE)
  1.3 ml L-glutamine, 200 mM (GIBCO #320-5039AE); store at 40°C for less than 1 month

Hanks balanced salt solution, pH 7.2 (GIBCO #310-40206)

Heparin (5000 international units (IU) per ml stock solution) or heparinized tubes (green top), without preservative

KCl, 0.075 M; 0.56 g in 100 ml distilled H_2O (hypotonic solution)

Liquid nitrogen

Methanol, HPLC grade (Fisher #A-452SK-4)

Phytohemagglutinin M, stock solution in sterile distilled water as recommended (GIBCO #R15-0576); after reconstitution, store frozen for several months

Ringer's solution (Sigma #K-4002)

Siliconizing solution—5% dimethyl-dichlorosilane (Sigma D-3879) in methylene chloride

Thymidine—stock solution (Sigma #T5018)

Trypsin—stock solution; 0.05% trypsin, 0.53 mM EDTA (GIBCO #610-5400 AG)
  For tissue culture: working solution—0.05% trypsin, 0.53 mM EDTA
  For GTG banding: 1 ml stock solution plus 49 ml Hanks buffered salt solution

Equipment

Cell scrapers (GIBCO #925-1160 YT)
Centrifuge, bench top
Coplin jars
Coverslips, 50 × 35 mm, #1 thickness (Fisher #12-545G)

Filter paper
Flasks, plastic tissue culture, 25 cm (Corning #25102)
Incubator, 37°C
Microscope, with phase-contrast objectives: 16× and 42×
Pasteur pipette, 9" glass (Fisher #13-678029C)
Petri dish, plastic, for tissue culture, 100 × 20 mm (Falcon #3003)
Razor blades
Tubes
    15 ml, graduated, conical, polypropylene (Falcon #2097)
    50 ml, graduated, conical, polypropylene (Falcon #2070)
Vortex mixer
Water bath, 37°C

Protocols

Protocol 1. Preparation of Chromosomes from Peripheral Blood T Lymphocytes (Whole Blood Microculture Method)

1. Draw venous blood into heparinized tube (green top) or into syringe containing 0.1 ml heparin stock solution (5000 IU/ml).

2. Add 5 drops heparinized blood (using sterile Pasteur pipette) to 5 ml pre-warmed chromosome 4A medium, or growth medium containing 0.2 ml phytohemagglutinin solution, in a 15-ml sterile conical polypropylene tube with screw top. Set up cultures in duplicate. Mix by gently inverting several times.

3. Place culture tubes on a slant (approximately 30°) in a 37°C incubator and leave for 72 hours.

4. Add 25 μl Colcemid solution to each tube (final concentration is 0.05 μg/ml) and mix by gentle inversion. Incubate for an additional 30 minutes.

5. Centrifuge at 800 g for 10 minutes and suction off supernatant.

6. Resuspend pellet in 5 ml 0.075 M KCl, previously warmed to 37°C. Thoroughly resuspend pellet by mixing using lowest setting of vortex mixer. Incubate at 37°C for 10 minutes.

7. Centrifuge at 800 g for 5 minutes and suction off supernatant.
8. Resuspend pellet in 5 ml cold 3 : 1 methanol : acetic acid by dropwise addition of fixative, tapping bottom of tube to resuspend pellet. Remove aggregates by gently drawing up solution in a Pasteur pipette.
9. Immediately centrifuge at 150 g for 2 minutes. Suction off supernatant.
10. Repeat Steps 8 and 9.
11. Resuspend pellet in 0.5 ml 3 : 1 methanol : acetic acid and place on ice.
12. Immediately drop 6–7 drops of fixed cells onto large (35 × 50 mm), cold, wet coverslips using a Pasteur pipette. The coverslips should be kept at −70°C for a few hours before use. Immediately before use, dip coverslips into ice-cold distilled water. Drop cells from a height of approximately 2–3 feet (arm's length) onto a slide held at a 30° angle. Immediately spread drop over the surface of the coverslip by blowing by mouth.
13. Examine first coverslip by phase microscopy to determine if spreads are at the correct density.
14. When the cells are nearly dry (few minutes), rinse off fixative by immersion in ice-cold distilled water.
15. Immediately place coverslips in 95% ethanol and store at −20°C.

Notes

1.1. As an alternative to using whole blood, use lymphocyte-rich plasma. Lymphocytes can be separated by allowing heparinized blood to settle. High yields of lymphocytes can be obtained from Ficoll–Hypaque gradient centrifugation as well (6–8). Lymphocytes can be cultivated and chromosomes can be prepared as described earlier.

1.2. Long-term lymphoblastoid suspension cultures can be harvested as described here. Colcemid is added during the rapid growth phase of the culture.

Protocol 2. Preparation of Chromosomes from Monolayer Tissue Culture Cell Lines

1. Seed cells at low density onto plastic Petri dishes.

2. Twenty-four hours after seeding, remove old medium and replace with fresh medium.

3. Forty-eight to 72 hours later, add 25 μl Colcemid solution per 5 ml medium and gently swirl. Return to 37°C CO_2 incubator for 10 minutes–1 hour, depending on cell line.

4. Remove cells from bottom of plate with cell scraper. Scrape into 15-ml sterile conical tubes. Centrifuge cells at 800 g for 10 minutes.

5. Suction off supernatant. Resuspend pellet in 5 ml pre-warmed 0.075 M KCl. A single-cell suspension should be made by pipetting cells up and down through a Pasteur pipette.

6. Incubate in a 37°C waterbath for 30 minutes.

7. Centrifuge cells at 150 g for 5 minutes. Suction off supernatant and resuspend cells in 3:1 methanol:acetic acid. Continue as described in Protocol 1, Step 8.

Notes

2.1. The total time in fixative should not exceed 10 minutes using this method.

2.2. Lower concentrations of fixative, as described by Korthof (15), may be used in the initial fixation step.

2.3. Higher yields of mitotic cells can be obtained by synchronizing cells with thymidine block (6, 9) or by using the "shake-off" method (10).

2.4. The optimum time for treatment of different cell lines with Colcemid, hypotonic solution, or fixative may need to be determined experimentally.

Protocol 3. Preparation of Chromosomes from Monolayer Cells Grown on Coverslips (6)

1. Seed cells onto sterile coverslips placed on the bottom of Petri dish. Coverslips should be of an appropriate size to fit into stage holder for microdissection (see Chapter 3). Cells should be seeded sparsely (approximately 20% confluence). Add appropriate amount of growth medium to adequately cover cells.

2. When cells are 40–50% confluent, gently remove medium and replenish with fresh growth medium.

3. Twenty-four hours later, add Colcemid to medium (final concentration should be between 0.04 and 0.2 µg/ml, depending on cell line). The optimum time to add Colcemid is when numerous mitotic cells are visible (rounded up) on coverslip.

4. Incubate cells at 37°C in Colcemid 1–4 hours. Then carefully draw off medium with a pipette to avoid removing mitotic cells. Slowly add pre-warmed (37°C) 0.075 M KCl to cover cells. Incubate in 37°C incubator for 30 minutes.

5. Add an equal volume of ice-cold 3:1 methanol:acetic acid without dislodging cells on coverslip. Leave in fixative 3–5 minutes.

6. Immediately remove coverslip with fine forceps and place in a Petri dish containing 95% ethanol at −20°C.

Notes

3.1. This method is particularly useful when working with slowly growing monolayer cultures.

3.2. The optimum times for the hypotonic and fixatives steps can be monitored with an inverted microscope. Using the shortest time in fixative that allows cells to rupture and chromosomes to spread open is preferable.

Protocol 4. Preparation of Chromosomes from Dipteran Salivary Glands (16, 17)

1. Salivary glands are dissected in insect Ringer's solution and are spread on a coverslip (35 × 50 mm) in a drop of 45% acetic acid (see Note 4.1). The coverslip should rest on a piece of filter paper.
2. The tissue is squashed by placing a siliconized coverslip on top of the tissue on the first coverslip.
3. The two coverslips are squeezed together, the thumb placed on top of the coverslip and the forefinger beneath the filter paper.
4. The two coverslips are held in place with forceps and quickly frozen in liquid nitrogen.
5. The coverslips are separated by flipping them apart with a razor blade.
6. The coverslip containing the squash is placed in three successive ethanol washes (70%, 95%, absolute) for 1 minute and either kept in 95% ethanol or air dried. The coverslip is kept in a moisture-free air-tight box until use.

Notes

4.1. As discussed earlier for metaphase chromosomes, chromosomes should not be exposed to fixative for more than a few minutes at room temperature.
4.2. Coverslips are siliconized by placing in siliconizing solution (see Reagents) for 5 minutes followed by one wash in 70% ethanol and four washes in distilled water.

Protocol 5. Solid Staining and GTG Banding of Metaphase Chromosomes (18)

1. Remove coverslips with chromosomes from ethanol and place in Coplin jar containing pH 6.8 buffer at room temperature.

2. For GTG banding, immediately remove coverslip from buffer and place in stock trypsin/EDTA solution. Incubate at room temperature for 1–3 minutes. This step is omitted in solid staining.

3. Place coverslips for 1 second in pH 6.8 buffer. Then place coverslips directly into Giemsa stain. Incubate at room temperature for 15 to 90 seconds depending on the intensity of the staining (see Note 5.1).

4. Wash twice in pH 6.8 buffer and air dry.

5. Perform microdissection immediately after drying.

Notes

5.1. The optimum time for trypsin treatment and staining should be determined experimentally.

5.2. Commercially supplied sterile trypsin solutions will not cause DNA degradation under the conditions described here.

5.3. If the microdissected DNA is to be used for biological experiments, sterile solutions (passed through bacteriological filters) and sterile Coplin jars must be used and work should be performed in a biological containment hood.

References

1. Mezzanotte, R., Vanni, R., Flore, O, Ferruci, L., and Sumner, A. T. (1988). Aging of fixed cytological preparations produces

degradation of chromosomal DNA. *Cytogenet. Cell Genet.* **48**, 60–62.

2. Darzynkiewicz, Z., Traganos, F., Carter, S., and Higgins, P. J. (1987). *In situ* factors affecting stability of the DNA helix in interphase nuclei and metaphase chromosomes. *Exp. Cell Res.* **172**, 168–179.

3. Stickel, S. K., and Clark, R. W. (1985). Mass characteristics of DNA obtained from chromosomes of a human carcinoma cell line. *Chromosoma* **92**, 234–241.

4. Tamm, C., Hodes, M. E., and Chargaff, E. (1953). The formation of apurinic acid from the deoxyribonucleic acid of calf thymus. *J. Biol. Chem.* **195**, 49–63.

5. Arrighi, F. E., Bergendahl, J., and Mandel, M. (1968). Isolation and characterization of DNA from fixed cells and tissues. *Exp. Cell Res.* **50**, 47–53.

6. Rooney, D. E., and Czepulkowski, B. H. (eds.) (1986). "Human Cytogenetics, A Practical Approach." IRL Press, Oxford.

7. Macgregor, H. C., and Varley, J. C. (eds.) (1988). "Working with Animal Chromosomes." Wiley, New York.

8. Verma, R. S., and Babu, A. (eds.) (1989). "Human Chromosomes. Manual of Basic Techniques." Pergamon, New York.

9. Stubblefield, E. (1968). Synchronization methods for mammalian cell cultures. *In* "Methods in Cell Physiology" (D. M. Prescott, ed.), pp. 25–43. Academic Press, New York.

10. Terasema, T., and Tolmach, L. J. (1963). Growth and nucleic acid synthesis in synchronously dividing populations of HeLa cells. *Exp. Cell Res.* **30**, 344–350.

11. Claussen, U. (1980). The pipette method: A new and rapid method for chromosome analysis in prenatal cells. *Hum. Genet.* **54**, 277–278.

12. Holmquist, G. (1979). The mechanism of C banding: Depurination and beta elimination. *Chromosoma* **72**, 203–224.

13. Iseki, S. (1986). DNA strand breaks in rat tissues as detected by *in situ* nick translation. *Exp. Cell Res.* **167**, 311–325.

14. Edström, J. E., Kaiser, R., and Röhme, D. (eds.) (1987). Microcloning of mammalian metaphase chromosomes. *In* "Methods in Enzymology," Vol. 151, pp. 503–516. Academic Press, New York.

15. Korthof, G. (1986). An improved fixation method for chromosome preparation of Chinese hamster–human hybrid and mouse cell lines. *Cytogenet. Cell Genet.* **41**, 181–184.

16. Pardue, M. L., and Gall, G. (1975). Nucleic acid hybridization to the DNA of cytological preparations. *In* "Methods in Cell Biology" (D. V. Prescott, ed.), pp. 1–16. Academic Press, New York.

17. Scalenghe, F., Turco, E., Edström, J. E., Pirotta, V., and Melli, L. (1981). Microdissection and cloning of DNA from a specific region of *Drosophila melanogaster* polytene chromosomes. *Chromosoma* **82**, 205–216.

18. Seabright, M. (1971). A rapid banding technique for human chromosomes. *Lancet* **ii**, 971–972.

Chapter 3

Methods of Chromosome Microdissection

Introduction

Physical dissection of metaphase chromosomes offers a direct approach to the cloning and analysis of DNA sequences that range up to thousands of kilobase pairs in length. The micromanipulation instruments (e.g., cutting microneedles and nanoliter volumetric pipettes) were developed by DeFonbrune (1). Using these techniques, Edström and his colleagues performed biochemical assays on microdissected cells and subcellular structures (2). Scalenghe et al. (3) and Henning et al. (4) then reported the application of these methods to chromosome microdissection experiments using *Drosophila* polytene chromosome spreads. Several reports followed that described microdissection of and cloning from metaphase mammalian chromosomes (5–11). Although most of these investigators used curved dissecting needles and inverted coverslips over an oil-filled chamber, we and others (11–13) find that using straight needles and slides that are placed upright on an inverted microscope is technically easier and more generally applicable. In this chapter, we describe the microinstruments and three methods that currently are in use for the microdissection of metaphase chromosomes. The first method uses an inverted microscope that is equipped with a video camera and TV display monitor. Dissection and micromanipulations are performed while watching on the display monitor, using the high magnification and image enhancement offered by the video camera. The second method uses an oil-filled chamber and a standard phase-contrast upright microscope. The third method uses an upright microscope that is equipped with a laser microbeam excimer. The chromosomes are cut with an attenuated laser beam. We prefer to use the video microscope method to prepare human chromosomal fragments.

Methods

Video Microscope Method

Microscope

The Diaphot inverted microscope we use for microdissection is one component of a "Chromosome Work Station" using components manufactured by Nikon, Inc., and Narishige Co., United States. An ideal microscope should have the following features. First, the microscope should have a short and efficient light path. Second, a focusable lens that permits rapid alignment of the phase annular rings should be incorporated in the turret assembly. Third, in addition to phase contrast, fluorescence and Nomarski differential interference contrast (DIC) systems are required for *in situ* hybridization and microinjection applications. These applications will be discussed in greater detail later in this manual. Fourth, several different specimen holders are required. These holders should be attached easily to the fixed stage and should remain stationary during focusing and other manipulations. The stage must be large enough to permit direct mounting of micromanipulators and provide sufficient working space. Fifth, a wide range of phase objectives ($4\times$, $10\times$, and $40\times$ or $60\times$) is important for the low and high magnification procedures. Another important feature in a microscope is a condenser that permits interchanging between the various optics (e.g., phase contrast, reflected light, and Nomarski) while maintaining parcentricity. Finally, for stability of the microscope, the entire optical system should be integrated into the stand completely. The Diaphot microscope (Nikon) offers the majority of these features. The microscope has a long working distance (LWD) condenser (working distance of 20.5 mm and numerical aperture of 0.52). The condenser has a phase annular ring for $4\times$, $10\times$, $20\times$, and $40\times$ phase objectives. Also, the Nomarski system uses the same condenser with a setting for the $10\times$ and $20\times$ lenses. Thus centering only one annular ring conveniently allows easy interchange between phase contrast and differential interference objectives, and ensures the control of parcentricity when changing magnifications. Several other microscopes such as the Axiovert 10 (Carl Zeiss, Inc., West Germany) and the Labovert FS (Wild Leitz USA, Inc.) offer similar features. The Axiovert microscope offers three additional features: the complete illumination system can be tilted back during specimen exchange; an optional motorized nosepiece changes quickly between objectives; and, finally, an image split system that permits the simultaneous viewing of the specimen on the video monitor and through the eyepieces. These features may increase equipment cost, however. For record keeping, a 35-mm camera and/or a video camera can be mounted readily on the built-in camera ports of the microscope. The positions of camera ports must not interfere with the various accessories on the microscope. The features of the Nikon Diaphot microscope are illustrated in Figure 3.1.

Video Camera, TV/Microscope Coupler, and Monitor

The coupling of a video camera to the microscope has become extremely useful for a wide variety of biological applications, particularly for viewing light microscopic images that are either contrast or light limited. Video microscopy permits structures to be seen with higher contrast magnification. With the phase microscope alone, resolving bands that correspond to a distance of as little as 10–30 Mbp along the chromosomal DNA is relatively difficult. Using the video microscope with its improved contrast capabilities, chromosomal fragments as small as 10 Mbp measuring about 0.4 μm can be resolved

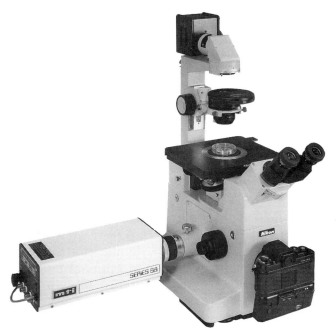

Figure 3.1 The Nikon Diaphot inverted microscope. This photograph shows the solid diecast body with low center of gravity and built-in binocular tube to ensure stability of the microscope. Also built into the microscope stand are twin camera ports for automatic 35-mm photography and closed-circuit TV (CCTV). The wide range of accessories that can be attached to this microscope are described in the text. (Reprinted with permission from Nikon Inc.)

readily. A video microscope system is also invaluable because of its real time capabilities that enable one to observe and record dynamic events—in this particular application, micromanipulations—as they occur. Video enhancement dramatically improves the quality of chromosomal spreads, enhances chromosomal banding, and permits visualization of secondary bands that normally are difficult to resolve.

Video cameras have an inherent magnification factor that ranges from 10 to 200×, depending on the make and quality of the camera. For example, the QX-104 video camera has an intrinsic magnification factor of 27× as calibrated using a microgrid (Graticules Ltd., England). Newly developed charge-coupled device (CCD) cameras such as the CCD-72 (Dage-MTI, Inc.) and the CCD-IRIS (Sony Corp.) are capable of yielding high resolution images and are extremely sensitive to low level light (10^{-5} lux). We use the CCD-72 camera series because of its compact size and light weight (Figure 3.2). This camera offers high image resolution, contrast enhancement features, and an impressive 40× magnification factor, as well as a selectable auto/manual operation, an adjustable edge enhancement mode, no geometric distortion, and a built-in gray-scale reference signal. The video camera magnification factor can be conditioned and increased by adjusting the electronic focusing of the camera. When this camera is coupled to our microscope, using a 40× or 63× objective lens, we easily can arrive at a magnification of ~1600–2500×. In addition, the use of high resolution TV/video camera microscope couplers can offer 0.5–2.0× magnification. The HR200 series 2× coupler connects to our CCD-72 camera series and remarkably improves the parfocality and focusing adjustment of the camera. The combination of a 2× coupler with the CCD-72 camera yields a final magnification of ~3200–5000× when using 40× and 63× objective lenses, respectively. The resulting video-enhanced image is displayed on a high resolution Electrohome ECM 1312 or equivalent VGA monitor (monitors have an additional magnification factor). Note that, although the magnification offered by the video camera is called empty magnification (i.e., magnification of a blurred image), the quality of the image can be improved greatly by the choice of the following accessor-

Figure 3.2 Dage-MTI camera (Model 72s). This camera is a charge-coupled device that is small, light, and sensitive to low light levels. The camera also features internal magnification factors and yields high resolution images.

ies: (1) high resolution microscope coupler, (2) interference contrast filter, (3) high numerical aperture lens (NA 0.8–1.4), and (4) high resolution display monitor, for example, super VGA 1024 × 768. To increase the resolution of this system further, an image intensification pickup system (e.g., Video Scope International, Ltd.; Model KS-1380) can be attached to the camera. The intensifier tube is adjusted for high gain, uniform output brightness, and linear amplification of the input light. The KS-1380, as are other intensifiers, is designed for use in applications in which the intensity and location of low light level objects are required.

Still pictures (at 3200–5000×) can be captured from the image monitor with the arrangement shown in Figure 3.3. A standard 35-mm camera or instant camera format is mounted on a sturdy support or hood in front of the monitor faceplate until the image fills the camera format. The camera is loaded with a moderately fast, fine-grain film (e.g., Kodak plus X or Tmax pan; 400 ASA) and the shutter speed is set to 1/8 (or 1/4) second. Exposures that are too short cause the upper or lower part of the photograph to appear noticeably darker, because of exposure to fewer cycles of the video scan than the rest of the film. In contrast, exposures that are too long result in a blurred image. When copying from the monitor, the room must be darkened or a hood must be used to eliminate scattering of the room light in the phosphor, which will reduce contrast. Also, reflections off the monitor faceplate may appear in the photograph.

To overcome the problems of taking pictures directly from the monitor, a video graphic printer can be used. Black and white video printers such as Sony Model UP910 (shown in Figure 3.4) are suitable for this application. These printers (also available in color, e.g., Model UP930) are fully compatible with a CCD-72 camera and allow instant recordings (5 × 7 format) of desired chromosomal spreads before and after dissection directly from the image monitor at the final magnification. Obtaining a photograph of a spread at the final magnification with adjusted contrast and resolution is more advantageous than photographing at 100X magnification and enlarging prints, particularly when microdissection is performed on a derivative chromosome from a cancer cell line or from a locus genetically linked to an

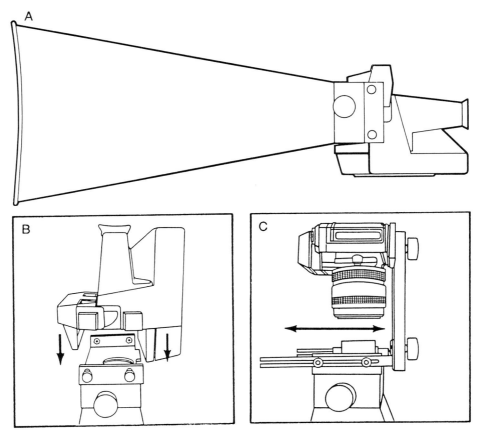

Figure 3.3 Schematics of instant recording system from the image monitor. This simple and affordable system permits the recording of images displayed on CRT monitors using Polaroid or 35-mm cameras. The system consists of a CRT hood that places the camera at the correct distance from the monitor and also shields the camera from light in the room (A). Two brackets are provided that allow the mounting of Polaroid cameras and 35-mm cameras (B and C, respectively). The Polaroid One-Step Land camera and Polaroid 600-Plus film are used. For 35-mm photography, use a shutter speed of 1/15 seconds or longer and medium aperture (f/5.6 to f/11) to ensure adequate depth of field. The monitor contrast and brightness are set at "normal" for most photography. (Adapted with permission from NPC Photo Division, Newton, Massachusetts.)

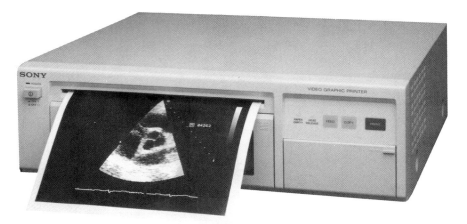

Figure 3.4 Black and white video graphic printer (Model UP910; Sony Corp.). This printer is easily adaptable to microscopes equipped with a video camera. Instant photos can be obtained from the image monitor (5 × 7 format).

inheritable disorder. In such cases, precisely recognizing the dissected region is crucial. Photographs obtained from the printer are suitable also for karyotyping and *in situ* hybridization experiments, which are discussed in later chapters of this manual.

Dissection Chamber

We designed a new circular rotating stage that permits the simultaneous microdissection and collection of chromosomal DNA fragments. The stage is a rectangular plate made of light aluminum (110 × 120 × 3 mm) in which a circular opening (90-mm diameter with a 1.0-mm edge) is made. The plate is mounted readily on the standard X–Y arms of the microscope stage. The circular slide holder contains two openings of 50 × 35 and 25 × 25 mm with a 1.0-mm edge to support the dissection and DNA collection slides, respectively. The slide holder can be dropped in easily or removed from the opening on the X–Y stage. The metaphase spreads are dropped onto a

50 × 35 × 0.17-mm (#1, Fisher Scientific) coverslip. The coverslip is placed in the 50 × 35-mm opening (chromosome spread side up). Thus, the spreads remain accessible from the top with a micromanipulator and microneedles. The underside of the coverslip is in contact with a high magnification lens (40× or 63×; NA 0.75–1.2). Since the spreads are accessible from above, better resolution of chromosomal bands can be achieved using the highest numerical aperture lenses (e.g., 1.2) without concern about altering the working distance for performing microdissection. Rotating the circular stage, the chromosome of interest can always be positioned perpendicular to the dissection needle tip. The microinjection slide is cut from a standard depression slide (VWR) to 25 × 25 mm to fit in the second opening. Figure 3.5 depicts the circular microdissection/DNA collection stage mounted on a standard microscope stage.

Micromanipulator

One or two micromanipulators are required for microdissection experiments. Normally, the right-hand manipulator is used for dissection of chromosomes. The other manipulator is used for manipulation of the chromosomal fragments (e.g., collection, transfer, and introduction into aqueous drops). Three-dimensional mechanical or hydraulic micromanipulators are best suited for this application. A hydraulic drive system ensures smooth vibration-free operation. The three-axis motion control also is required to achieve extremely high precision movement. In addition to its x, y, and z axis movement, a micromanipulator must provide a single direction angular movement. The Leitz ultrafine mechanical micromanipulator allows this type of mechanical movement in three dimensions. This manipulator also can be adjusted with coarse and fine screw controls that are useful for proper positioning of the instrument that holds the needle tip within micrometer distances of the chromosome. Since keeping vibration to an absolute minimum is essential, a fine drive joystick is used to reduce hand movement during dissection, separation, and collection of fragments. The features of the Leitz micromanipulator are illustrated in Figure 3.6. The Narishige manipulator Models MN-

Figure 3.5 The top view of the circular microdissection/DNA microinjection stage. The design of this stage permits the simultaneous microdissection and collection or microinjection of chromosome fragments. The rotating stage also allows the dissection of any chromosome on the slide, irrespective of its orientation. The stage is made of a rectangular plate of light aluminum (110 × 120 × 3 mm). The circular disk (90-mm diameter) rotates freely and contains two openings to hold the 35 × 50-mm dissection slide and the 25 × 25-mm collection/injection slide. The plate is mounted on the standard X–Y stage of the microscope. See Figure 3.13b for illustration.

162 and MN-163 (Narishige) also provide ultrafine mechanical joystick movement and are a favorable alternative to the Leitz instruments. In addition to permitting extremely small movements, these instruments have a rather compact design and can be clamped directly to the microscope stage, thus minimizing vibrations.

Vibration-Free Tables

The microscope and the micromanipulators are placed on a tabletop isolation platform (e.g., Vibra-Plan, Model 2208, Kinetic Systems,

Inc., Figure 3.7). This low-cost, passive air-mount template has front dual air-inflation control. When inflated to a load-supporting pressure, the platform floats freely and the system is set automatically at the correct operational pressure, which ensures that no vibrations are transmitted to the equipment. Other tables (e.g., marble tables)

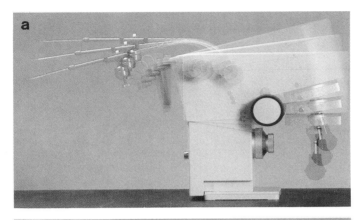

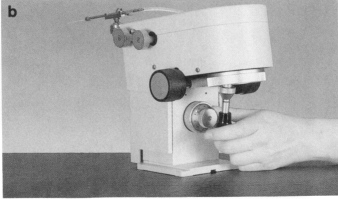

Figure 3.6 (a) Leitz ultrafine micromanipulator. This right-hand manipulator device consists mainly of a lever and a slide rail mechanism, which converts the coarse movement of the hand into a very delicate micromovement. (b) The manipulator permits movements in vertical, horizontal, forward, and reverse directions as well as tilt adjustment. A micropipette holder is also shown.

Figure 3.7 Kinetic Systems Vibra-Plan (Model 2208). This tabletop isolation platform floats freely on a bench when filled with air, minimizing vibrations in the microdissection apparatus.

may be adequate to dampen vibrations, depending on the vibration noise in the room and the building in which the experiments are being performed.

Needles, Needle Holders, and Needle Pullers

Borosilicate capillary tubes (with or without inner filaments) with 1-mm o.d. × 0.65-mm i.d. [World Precision Instruments (WPI), Stoelting and Diamond General Corp.] are best suited for making microneedles and micropipettes. Several micropipette holders are commercially available that fit in most micromanipulators. Microinjection needle holders from Eppendorf, Leitz, and Narishige are commonly used. Also, several needle pullers are available to help fashion various cutting needles, holding pipettes, and volumetric pipettes. We have tested most commercially available needle pullers and have found that the Narishige PP-83 two-stage needle puller is cost effective and easy to use, and consistently pulls needles with a specifically programmed tip diameter. Other micropipette pullers are available from Sutter Instruments Co. (Model p-87) and Technical Products International, Inc. (TPI). Sutter Instruments also offers a micropipette beveler (Model BV-10) that helps fashion special pipette tip configurations. Figure 3.8 shows the microneedle puller Model PB-7 (Narishige) which is identical to Model PP-83.

Pulling Dissection Needles and Micropipettes

Dissecting needles and nanoliter transfer pipettes which are drawn to a 1- to 2-mm long taper and tip diameter of less than 0.5 μm are flexible enough to permit pressure on the tip of the needle without breaking. The first-stage heating element on the Narishige two-stage

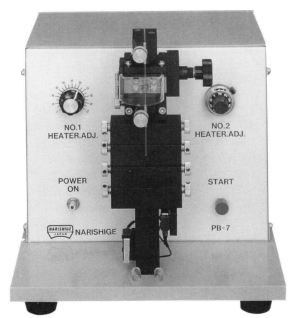

Figure 3.8 The Narishige pipette puller (Model PB-7) can produce pipette tips of less than 0.1 μm with a single pull and pipette tips from a few micrometers to tens of micrometers in the double-pull mode. The wide range of pipette tip sizes is provided by combinations of weights and two heater controls. These controls are very sensitive to small voltage changes. The PP-83 model we use is identical to the PB-7, except it has a constant weight.

puller is turned on to melt a 6-inch capillary until a constriction of about 5.0 mm is made in the middle of the capillary. The heating filament is then moved downward around the constriction and the second-stage heater is switched on until the glass separates into two pipettes. These steps are illustrated in Figure 3.9. The balance between needle flexibility and sharpness must be determined empirically by the experimenter. The balance depends on the glass and the type of pipette pullers used, on the temperature used to melt the glass, and on the tension applied to the capillary.

After micropipettes are pulled, they can be fashioned for use as dissection needles or other microtools with a microforge (Figure 3.10) or a micropipette grinder (Figure 3.11). For example, a microforge allows the preparation of curved pipette tips, hooks, or round tips using fire polishing. Beveled pipette tips can be prepared with the grinder.

Measuring the Tip Diameter of Needles

The internal tip diameter may be estimated using an interval measuring grid (e.g., Graticules LTD., England). These grids usually have 1- to 2-μm gradations. We developed a simple and precise method to determine a pipette inner tip diameter as small as 0.2 μm (14). The method is based on the Laplace equation. Briefly, the needle is submerged into methanol or water and air pressure is applied through the back end of the pipette until a stream of bubbles is observed. The bubble pressure is related inversely to the pipette tip inner diameter and the surface tension of the liquid. The bubble diameter measured by this method is in excellent agreement with that measured directly by the scanning electron microscope. In some microdissection experiments, however, the tip diameter does not have to be measured precisely.

Delivery and Suction System

The dissecting needle is fitted into an Eppendorf microinjector holder. The micropipettes are connected with Tygon or Teflon tubing to

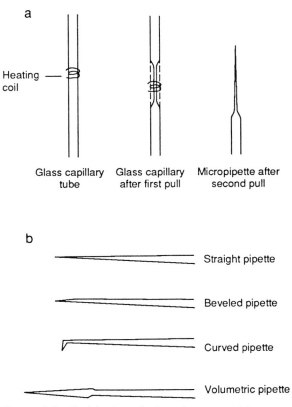

Figure 3.9 Fashioning of micropipettes. (a) Two-stage pulling of a glass micropipette. (b) Examples of fashioned micropipette tips. A straight pipette tip is used for microdissection on inverted microscopes. The curved tip is made using a microforge (Figure 3.10) and is particularly useful for microdissection using the oil chamber method. A beveled pipette tip is useful in microinjection of cells or ova, minimizing cell injury. Beveled tips are prepared using a micropipette grinder (Figure 3.11). The volumetric pipette is used to transfer nanoliter volumes of solutions.

a glass syringe connected with a luer-lock attachment to a timer/motorized electric microinjector (Narishige IM-1 system) that allows the delivery and suction of nanoliter volumes. Other systems that allow delivery and suction of nanoliter volumes (e.g., Eppendorf

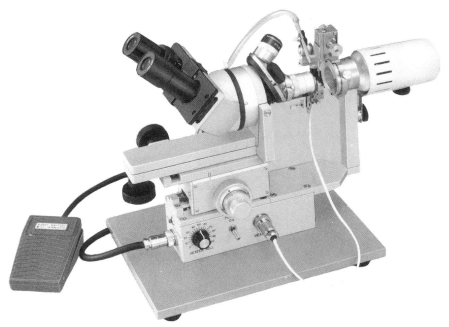

Figure 3.10 Micropipette microforge (Narishige Model MF-9). This instrument smoothes and rounds uneven sharp pipette tips by fire polishing and facilitates production of microtools (e.g., curved tips or hooks).

Microinjection System 5242 or homemade systems using micrometer hand-driven syringes, e.g., Stoelting Model 5222) may suffice. Figure 3.12 shows the Eppendorf microinjector. The entire system (needle, tubing, and syringe) is filled with silicone fluid (Corning #705) which we have found to be inert for most biochemical procedures. Spectroscopic grade paraffin oil is equally useful. When the chromosomal fragment dissected is larger than 0.5 μm, a relatively larger pipette can be used for dissection as well as for collection of the fragment. Figure 3.13 illustrates the components of the complete video microscope apparatus.

Methods 57

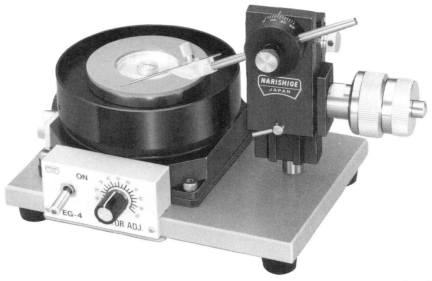

Figure 3.11 Micropipette grinder (Narishige Model EG-4). This instrument bevels the pipette to increase opening diameter while retaining the small size of tip.

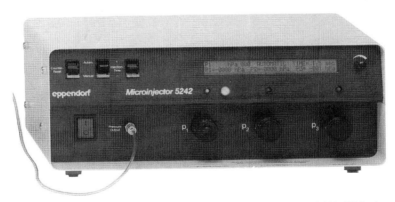

Figure 3.12 Eppendorf microinjector (Model ECET 5242). This instrument is equipped with switch controls that provide injection pressure, holding pressure, and negative pressure (vacuum). The vacuum mode is used to fill micropipettes. The instrument operates using compressed gas (air or nitrogen).

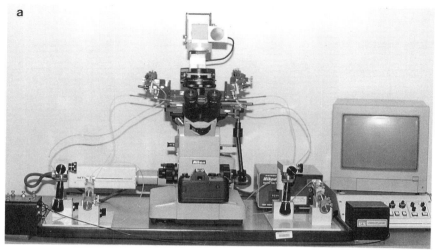

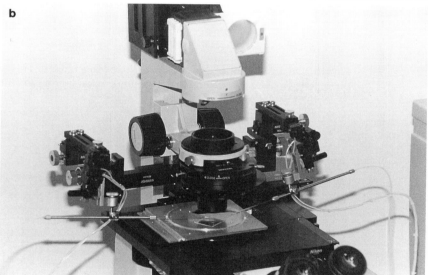

Figure 3.13 (a) Complete video microscope microdissection apparatus. This system is composed of the Nikon Diaphot microscope, video camera, high-resolution display monitor with gray levels and contrast controls, two micromanipulators, two microsyringes, and the rotating microdissection/microinjection stage. Other accessories include micropipettes and pipette holders, 35-mm camera and epifluorescence excitation filters, and power supply. The entire system is placed on a table-top vibration-free system (Kinetics System, USA; see Figure 3.7). (b) A close-up photograph illustrating the rotating stage and the micromanipulators mounted on the microscope stage.

Slides for Collecting Chromosome Fragments

Chromosome fragments are pooled into an aqueous drop 1–2 nl in volume. To avoid evaporation, the collection droplet is introduced into a small volume of silicone or paraffin oil (200–500 μl) on a slide with a 0.8-mm circular depression (VWR #48333-002) that is placed next to the dissection slide (Figure 3.14). Alternatively, the aqueous drop could be placed under a moist chamber on the side of the dissecting chamber (11). For subsequent injection or cloning experiments, the dissected fragments are pooled in a drop of Tris–HCl buffer (pH 7.2) containing 2 mM EDTA, 500 μg/ml proteinase K, and 0.05% sodium dodecyl sulfate (SDS). Avoid proteinase K/SDS if subsequent PCR will be used. The extraction DNA drop is then transferred to a microcentrifuge tube. A 1-μl supply drop of buffer is introduced under oil using an Eppendorf pipetter. The nanoliter drops are prepared using a fine micropipette and suction, and placed near the supply drop for easier relocation after each dissection. The relative size of the nanoliter drop to that of the supply drop is illustrated in Figure 3.14.

Microdissection and Fragment Collection

The use of a video camera and high resolution monitor coupled to the Diaphot microscope increases the precision of cutting a specific chromosomal band. Using the system described here, cutting chromosome fragments of less than 0.5 μm in length is possible. The precision of the dissection is limited by the degree of magnification and the tip size of the dissecting needle. However, isolating single minor bands of human metaphase chromosomes is possible.

Coverslips containing the metaphase spreads are secured into the slide holder, spread side up, and mounted onto the microscope mechanical rotating stage. Metaphase chromosomes are prepared as described in Chapter 2 and are dropped on thoroughly cleaned 50 × 35 × 0.17-mm (#1, Fisher Scientific) coverslips. The spreads are stained, then examined at 200× magnification to locate well-spread chromosome sets. The X and Y coordinates of qualified spreads are recorded using the microgrid on the mechanical stage.

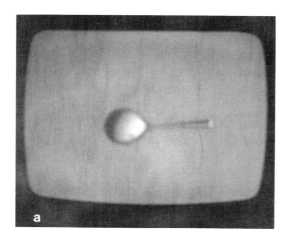
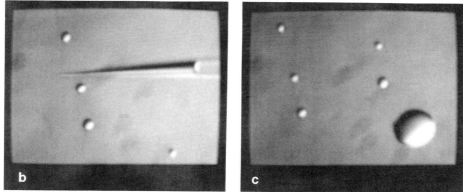

Figure 3.14 Preparation of nanoliter microdrops for collection and extraction of chromosomal DNA fragments. A micropipette (1-μm tip diameter) is filled first from the back end with oil using a 27-gauge spinal needle. Place 1 μl aqueous solution under oil. Place the micropipette under oil and fill with buffer from the tip using suction. (a) Transfer 15–20 nl supply drop. (b) Several nanoliter drops are introduced nearby to be used later for collection and resuspension of chromosome fragments. (c) Relative size of the microdrops to the supply drop. These operations are performed at 10× magnification (about 800× on the display monitor). Photographs were recorded directly from the monitor using the CRT-hood assembly described in Figure 3.3.

The desired chromosome is positioned perpendicular to the dissection needle. Suitable chromosomes are placed in the correct orientation by rotating the stage, using the same magnification. The dissection needle is brought into the field within a few micrometers of the spread. The system is switched to the high magnification lens (40× or 63×), and the video camera and monitor are used to perform microdissection.

Microdissection is performed by moving the capillary tip across the chromosome band. The dissected region will detach from the slide and should be pushed away from the chromosome. Successful dissections are verified by a clear zone in the chromosome arm where the cut was made. Fragments that may become contaminated by touching other chromosomal material are not collected. Recovery of the chromosome fragment is best done by applying some silicone oil onto the fragment through the dissection needle, then using suction of the micrometer syringe to secure the fragment to the tip of the needle under the closed system. Alternatively, to avoid the use of oil, the fragment is rehydrated with a buffer solution [20 mM Tris–HCl (pH 7.2), 100 mM NaCl, 2 mM EDTA] to make the fragment adhere to the tip of the capillary more securely. Dipping the needle in ethanol and drying in air also aids in adhering fragments to the needle tip. The suction method, however, ensures that the fragment will not be lost when removing the capillary from the slide to the collection drop, especially when breaking the surface tension between oil and air. The chromosome fragment on the tip of the dissection needle is expelled into the 1- to 2-nl drop under oil in a depression slide by applying positive flow using the microinjector. Dissections are made only on spreads in which the chromosomes are separated sufficiently, preferably when the chromosome of interest is located in the periphery of the metaphase spread. This location is particularly important when dissecting chromosomes in complex karyotypes of cancer cells. The process of cutting, collecting, and transferring a single fragment is completed in approximately 5 minutes. Typical dissections of major and minor chromosome bands are shown in Figure 3.15.

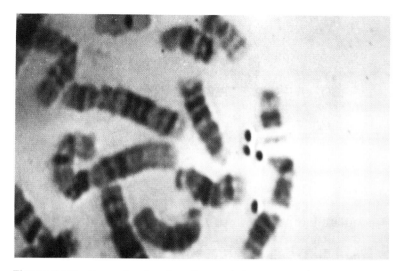

Figure 3.15 Example of a microdissected chromosome fragment. Human chromosome 3 (prepared from peripheral blood lymphocytes) is dissected at 40× (3200× on the display monitor). The fragments dissected, from top to bottom, are from bands 3p21, 3p14, 3p13, and 3q21. Each fragment measures about 0.5 μm and contains 10–15 Mbp. This photograph was taken directly from the display monitor to illustrate the magnification and chromosomal banding resolution.

Advantages of the Video Microscope Method

The video microscope dissection system shown in Figure 3.13 offers the following advantages:

1. The inverted microscope permits a convenient working distance for performing all subsequent micromanipulations.

2. The system provides three- to fivefold magnification over existing methods.

3. The system overcomes light- and contrast-limited situations with image enhancement provided by the video camera and monitor.

4. Instant photographs and/or video prints from the display monitor can be taken at each stage of microdissection, or the process can be recorded on videotape.

5. Enhanced video magnification permits higher precision in dissecting smaller chromosomal bands. For example, chromosome 1 (~7 μm) measures ~2.8 mm on the microscope ocular at 40× magnification, whereas it measures 2.24 cm on the display monitor. Thus, an average chromosome GTG band of 0.4 μm will measure ~2.0 mm.

6. Essentially, microdissection is performed in air where visualizing the tip of the dissecting needle and the chromosome fragment is relatively easy. Fragment collection is performed under oil, however.

7. Fragment loss during transfer is avoided by applying suction at the pipette tip.

8. Straight dissecting needles and collection pipettes are prepared easily with the help of a simple needle puller.

9. Chromosomes can be dissected precisely by placing them perpendicular to the dissecting needle. The compound dissection–injection stage that we designed allows complete rotation of the spread so the desired chromosome can be dissected irrespective of its position in the metaphase spread. This new stage design also increases the number of chromosomes qualified for microdissection from a single slide.

A similar and relatively simpler system that uses an inverted microscope and avoids the use of the oil chamber has been described also (11).

The time required to dissect a specific chromosome region depends on quality and number of metaphase spreads on the coverslip. Usually 10–15 fragments can be dissected and pooled in a microdrop in 1 hour. About 80 fragments are required for cloning; fewer are required if the polymerase chain reaction (PCR) is used to amplify dissected DNA (see Chapter 4).

Troubleshooting

1. *Poor chromosome resolution.* Use orange interference filters (chromosomes are stained blue). Adjust monitor contrast, intensity of light source, or camera gray levels.

2. *Chromosomes not well spread.* Refer to Chapter 2 for appropriate timing of hypotonic and fixative steps and for "smashing" spreads on coverslips.

3. *Chromosome fragment cannot be picked up by needle.* Check closed, oil-filled system for leaks. Too much cytoplasmic debris is on coverslips (needs longer fixation). Fragment is too dry (can be rehydrated by placing in a moist chamber for a few minutes). Coverslips should be cleaned thoroughly before use.

4. *Poor visualization of the needle tip under oil.* Tip of needle is not filled with oil.

5. *Loss of fragment during transfer.* Tip surface tension or static charge; use siliconized pipettes and prewash with ethanol. Oil-filled system is not closed. Air bubble in system. Suction is not strong enough.

6. *Fragments repelled by needle tips.* Static charge on the tip or too much silanization of dissection slide or pipette; prewash slides and pipettes with ethanol and acid, followed by wash in excess distilled deionized water. Oil flowing from pipette tip; reverse oil flow from pipette.

7. *Microdrops do not stay at the bottom of the depression slide.* Surface is not clean or is oversilanized; wash with detergent and excess distilled water.

8. *Needle tip dulls or breaks after dissection.* If needle taper is too short it will break easily on contact with slide surface; use needle

with long taper to withstand pressure during fragment cutting (see section on needle pulling).

9. *Difficulty finding desired chromosomes.* Screen slide beforehand at low magnification and mark qualified spreads (determined by orientation, spacing between chromosomes, banding, etc.) with a pen (small dots); then use the rotating stage to position marked chromosomes at high power.

Oil Chamber Method

The oil chamber method was first used for dissecting and microcloning polytene chromosomes by Scalenghe and colleagues (3) and has been applied to the dissection of metaphase chromosomes (5–11). This technique is discussed in detail in several methods papers (3, 5). The oil chamber consists of a one-piece glass slide that measures 70 × 35 × 6 mm with a rectangular groove, 25 mm wide and 3 mm deep, and 12 × 30 × 0.17-mm glass coverslips on which chromosomes are prepared (see Figure 3.16). Glass chambers and coverslips

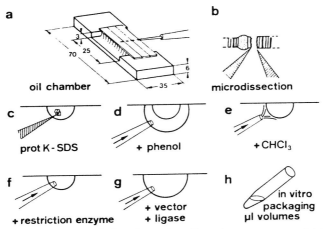

Figure 3.16 Oil chamber microdissection apparatus. (a) Dimensions of the oil chamber (in mm). (b–h) The steps of microcloning operations. [Reprinted with permission from Pirotta *et al.* (27).]

are washed first in hydrochloric acid, then in distilled water before use. In this method, an upright phase-contrast microscope equipped with an LWD condenser to view through the 6-mm thick oil chamber is used. All other tools described in the previous section, for example, manipulators, micrometers, needle puller, optical filters, and vibration-free surfaces, are suitable for the oil-chamber method as well, although curved dissecting needles (fashioned using a microforge) must be used.

The coverslip containing the chromosomes is placed on the groove (chromosome spread side down). The space between the coverslip and the bottom of the chamber is filled with paraffin oil. Microdissection is performed on the lower surface of the coverslip using a dissection needle with a tip that is bent in a microforge to an angle of ~45° from the axis. Dissection is performed under a high-power dry objective (at least 40×). All subsequent manipulations, such as micropipetting and transfer of DNA fragments, are done with low-power objectives (10× and 20×) using various nanoliter volumetric pipettes. To remove a chromosome fragment, the tip of the needle is first moistened in a droplet of glycerol/phosphate buffer [4 volumes 87% glycerol and 1 volume 0.05 M sodium–potassium–phosphate buffer (pH 6.8)] (3). The chromosome is cut across with the pipette tip and dragged through the liquid paraffin into the collection drop of glycerol/phosphate buffer. The chromosome fragment can be freed from the tip with a second needle if necessary. After collecting several fragments, the collection drop volume is reduced by suctioning with a micropipette. The remaining chromosome aggregate is combined with a separate microdrop of proteinase K/SDS solution [500 μg/ml proteinase K; 10 mM sodium chloride; 0.1% SDS; 10 mM Tris–HCl (pH 7.5)] and incubated at 37°C for 2 hours. The DNA is extracted three times by adding 4 nl phenol to the aqueous drop. The DNA is extracted further with chloroform to remove the phenol. The extracted DNA can be digested with restriction enzymes for further cloning using standard procedures. The oil chamber and a summary of the microcloning operations are shown in Figure 3.16.

Laser Microdissection Method

The blue-green argon ion laser microbeam was introduced as a tool for subcellular microsurgery in the late 1960s (15, 16). Work with the blue-green laser led to the development of a tunable wavelength flash-lamp pumped dye laser microbeam (17) and, later, to the low-power neodynium-YAG (yttrium–aluminum–garnet) dye laser (18). The use of dye laser microbeams in biology, for example, in microirradiation of nucleolar regions and microdissection of mitotic and cytoplasmic organelles, has been reviewed by Berns *et al.* (19). Microdissections of metaphase chromosomes of human lymphocytes and telomeric sequences from polytene chromosomes of *Drosophila melanogaster* were performed using a UV laser coupled to an inverted microscope (20, 21). Laser microdissection was used to generate libraries of clones corresponding to regions on the fragile X site, the Huntington disease on chromosome 4, and a chromosomal translocation between chromosomes 1 and 7 (22–24). In contrast to the two mechanical methods previously described, this optical method is based on the fact that optically active biological material can break down or melt on irradiation with light at very high photon densities (22, 23). Thus, even the chemical bonds are cleaved when biological material is heated locally to a few thousand degrees for a time ranging from nanoseconds to microseconds. The damage often can be confined to a specific cellular or subcellular target in a consistent and controllable way.

A typical laser microbeam apparatus, such as the one described by Monajembashi *et al.* (20), consists of an excimer laser as a primary source of laser light (e.g., Lambda Physik EMG 103 MSC). The microbeam quality can be improved by selecting a specific excitation wavelength. Therefore, an additional dye laser with a tunable wavelength (217–800 nm) normally is required (e.g., Lambda Physik FL 2002). A circular disk system of concentric rings is used to deliver high pulse energies at a repetition rate of about 10 Hz. By choosing a ring with a radial distance corresponding to the wavelength in use, energy densities of more than 10^{14} W/cm^2 can be achieved. The pulses are directed into an inverted microscope via the fluorescence illumination path and focused through the objective into the object plane. The pulses of laser, 20 nanoseconds in length, are directed into the microscope using optics similar to those of a fluorescence microscope (e.g., Ultrafluar 100, NA 0.85; Zeiss, Germany). By focusing through the objective lens, the pulses can be focused to the limits of diffraction.

This system provides pulse energies above 1 mJ and exposure durations as short as 25×10^{-12} seconds at a wide range of wavelengths between 320 and 800 nm, and up to 10 mJ at optimum wavelength for several dyes (e.g., neodynium-YAG). A UV laser is particularly suitable for dissecting chromosomal DNA because the laser damaging effect is limited to nanometers from the cutting region. Therefore, secondary damage is negligible. Also, since the DNA contained in chromosome slices absorbs light at wavelengths below 300 nm, microdissection using a UV laser above 300 nm will retain the biological integrity of the DNA and its suitability for subsequent microcloning procedures.

Chromosomal spreads of human lymphocytes are prepared by standard procedures on a coverslip. The chromosome is cut into slices by the diffraction ring system of a single laser pulse. The thickness of each slice is about 0.5 μm. Then the pulse is attenuated so the ring system no longer causes lesions in the chromosome. The central disk is used to destroy all parts of the chromosome except the slice needed for microcloning. For this procedure, homogeneously stained or unstained chromosomes are used and equidistant slices along the chromosome are cut. Obtaining a single slice of a chromosome requires ~2 minutes. To obtain a specific band of a GTG-stained chromosome, however, only the central disk of an attenuated beam can be used. This process can take significantly longer (tens of minutes). In either case, the slice can be taken up by a micropipette and placed into a microdrop, to be used as described for other microdissection methods. The microbeam dissection apparatus is potentially useful in generating equidistant and sequential chromosomal slices suitable for region-specific chromosomal libraries. However, in addition to the optical knife, all other tools necessary to manipulate the cut fragments, for example, micromanipulators, micropipettes, and microdrops, are still required. Although this technology offers a greater accuracy in obtaining extremely small fragments, relatively expensive equipment and very skilled hands are required. (See Figure 3.17 for schematics of the laser microbeam dissection apparatus.)

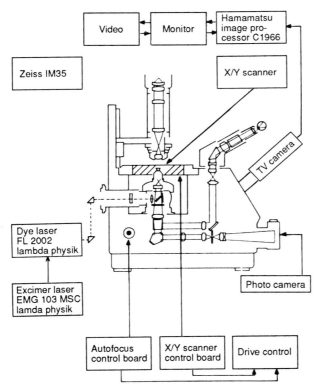

Figure 3.17 Diagram of the laser microbeam microdissection apparatus. The various components of the microscope are shown. Although microinstruments that are necessary for fragment collection and manipulation are not depicted here, they are essentially the same as described for the video microscope method. [Reprinted with permission from Ponelies *et al.* (21).]

Summary of Chromosome Microdissection and Collection for DNA Cloning

Reagents and Equipment

Buffer—20 mM Tris–HCl (pH 7.2), 2 mM EDTA, 100 mM NaCl
Capillary tubes (6 in. long, 1-mm o.d., 0.65-mm i.d.)
Microdissection and collection slides, sterile and precleaned
Micromanipulators
Micropipette puller (e.g., Narishige PP-83 or TPI)
Microscope equipped with video camera and monitor. Choice of accessories attached to microscope may vary from one microdissection method to another.
Microsyringes, tubing, and micropipette holders
Proteinase K (10 mg/ml), sterile DNase-free stock solution
Silicone or paraffin oil, spectroscopic grade (filter sterilized)
SDS, 2% stock solution
Vibration-free surface

Protocol

A summary of the steps required to perform a successful microdissection experiment is presented here. Some steps and instruments may vary depending on the choice of method for microdissection.

Before Microdissection

1. Prepare cells with high mitotic index according to standard procedures described in Chapter 2.

2. Prepare GTG-banded chromosomes immediately before microdissection.

3. Prepare volumetric micropipettes (i.d. ~1–2 μm) and straight microdissection needles (o.d. = 0.2–0.5 μm) using precalibrated settings on the micropipette puller. Fill both pipettes with oil from the back end using a 27-gauge spinal needle.

4. Place 200 μl filter-sterilized paraffin oil on the depression (fragment-collection) slide.

5. Fill a volumetric pipette with proteinase K/SDS solution (final concentration is 500 μg/ml proteinase K, 0.05% SDS) by dipping its tip in 1-μl supply drop placed under oil and applying gentle suction. Transfer a 10-nl drop of proteinase K solution to nearby area. This drop will be used to extract chromosomal DNA. Avoid proteinase K/SDS if fragments are to be subjected to PCR.

6. Similarly, place several 1- to 2-nl drops of aqueous Tris–HCl/EDTA buffer adjacent to the proteinase K drop using another volumetric pipette. The aqueous droplets are used later for washing and resuspending extracted DNA. Steps 5 and 6 are performed at 10× magnification.

Microdissection and Fragment Collection

1. Place the microdissection slide containing the chromosome spreads on the rotating stage. Find qualified spread(s) at 10× magnification and mark with a marking pen.

2. Use the rotating stage to orient the chromosome of interest perpendicular to the incoming microdissection needle.

3. Place the dissecting needle within a few micrometers of the spread at 10× magnification.

4. Switch the system to high magnification (40–60×) and video display (3200–5000×).

5. Approach the chromosomal band to be dissected at an angle of 30–45°. Using the micropipette, apply a few drops of oil to cover the chromosome and provide a closed system during the suction and collection of the fragment. Alternatively, the oil may be applied after dissection of the fragment.

6. Scrape the fragment off the slide while applying suction on the microsyringe. We find that this procedure ensures the adherence of the fragment to the tip of the micropipette.

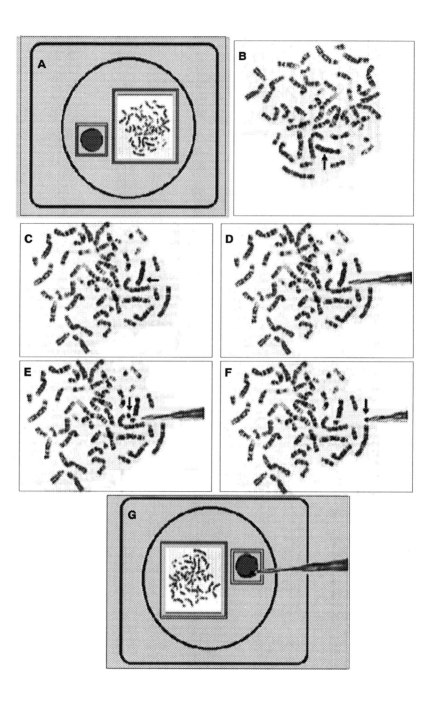

7. Apply further suction to secure the fragment firmly to the pipette. Hold the suction during transfer.

8. Switch the microscope to low magnification and bring the depression slide (containing DNA collection and extraction drops) into the field of view.

9. Gently introduce the micropipette containing the chromosomal fragment under the oil and place its tip into the drop containing the proteinase K/SDS solution.

10. Expel the fragment into the drop by reversing the micrometer into the injection mode. (Apply gentle suction and ejection several times to ensure fragment transfer to the drop.) The fragment should be visible in the drop at this time.

11. Repeat Steps 1–10 until a sufficient number of fragments is collected.

The sequence of microdissection steps is summarized in Figure 3.18.

References

1. DeFonbrune, P. (1949). "Technique de Micromanipulation," Monographes de l'Institut Pasteur, Masson, Paris.

Figure 3.18 Sequential steps of chromosome microdissection using the video microscope method. (A) Locate a chromosome spread on the dissection slide at 10× magnification. (B) Shift to higher magnification to identify the target chromosome for microdissection. (C) Rotate slide to orient target chromosome perpendicular to the incoming dissection needle. (D) Lower the dissection micropipette and position the tip proximal to the target chromosomal band. (E) Cut the chromosome fragment by scraping across the band in one forward motion. (F) Apply suction to the pipette to collect and secure the fragment to the tip of the micropipette. (G) Lift the micropipette up slowly; rotate the stage to bring the depression slide in field of view. Lower the pipette, while holding suction, into the aqueous drop under the oil. Eject the fragment into the drop by switching to injection mode.

2. Edström, J. E. (1964). Microextraction and microelectrophoresis for determination and analysis of nucleic acids in isolated cellular units. *In* "Methods in Cell Physiology" (D. M. Prescott, ed.), pp. 417–444. Academic Press, New York.

3. Scalenghe, F., Turco, E., Edström, J.-E., Pirrotta, V., and Melli, M. L. (1981). Microdissection and cloning of DNA from a specific region of *Drosophila melanogaster* polytene chromosomes. *Chromosoma (Berlin)* **82**, 205–216.

4. Henning, W., Huijuser, P., Vogt, P., Jackle, H., and Edström, J.-E. (1983). Molecular cloning of microdissected lampbrush loop DNA sequences of *Drosophila hydei*. *EMBO J.* **2**, 1741–1746.

5. Röhme, D., Fox, H., Herrmann, B., Frischauf, A. N., Edström, J. E., Mains, P., Silver, L. M., and Leharch, H. (1984). Molecular clones of the mouse t complex derived from microdissected metaphase chromosomes. *Cell* **36**, 783–788.

6. Fisher, E. M. C., Cavanna, J. S., and Brown, S. D. M. (1985). Microdissection and microcloning of the mouse X chromosome. *Proc. Natl. Acad. Sci. USA* **82**, 5846–5949.

7. Bates, G. P., Wainwright, B. J., Williamson, R., and Brown, S. D. M. (1986). Microdissection and microcloning from the short arm of human chromosome 2. *Mol. Cell Biol.* **6**, 3826–3830.

8. Kaiser, R., Weber, J., Grzeschik, K.-H., and Edström, J.-E., *et al.* (1987). Microdissection and microcloning of the long arm of human chromosome 7. *Mol. Biol. Rep.* **12**, 3–6.

9. Weith, A., Winking, H., Brackmann, B., Boldyreff, B., and Traut, W. (1987). Microclones from a mouse germ line HSR detect amplification and complex rearrangements of DNA sequences. *EMBO J.* **6**, 1295–1300.

10. Martinsson, T., Weith, A., Cziepluch, C., and Schwab, M. (1989). Chromosome 1 deletions in human neuroblastoma: Generation and fine mapping of microclones from the distal 1p region. *Genes Chrom. Cancer* **1**, 67–78.

11. Senger, G., Lüdecke, H.-F., Horsthemke, B., and Claussen, U. (1990). Microdissection of banded human chromosomes. *Hum. Genet.* **84,** 507–511.

12. Wesley, C. S., Ben, M., Kreitman, M., Hagag, N., and Eanes, W. (1990). Cloning regions of the *Drosophila* genome by microdissection of polytene chromosome DNA and PCR with nonspecific primers. *Nucleic Acids Res.* **18,** 599–603.

13. Lüdecke, H.-F., Senger, G., Claussen, U., and Horsthemke, B. (1989). Cloning defined regions of the human genome by microdissection of banded chromosomes and enzyme amplification. *Nature (London)* **338,** 348–350.

14. Hagag, N., Randolf, K., Viola, M., and Lane, B. (1990). Precise, easy measurement of glass pipet tips for microinjection or electrophysiology. *Biotechniques* **9,** 401–406.

15. Berns, M. W., Olson, R. S., and Rounds, D. E. (1969). *In vitro* production of chromosomal lesions with an argon laser microbeam. *Nature (London)* **221,** 74–75.

16. Berns. M. W., and Rounds, D. E. (1970). Cell surgery by laser. *Sci. Am.* **222,** 98.

17. Berns, M. W. (1972). Partial cell irradiation with a tunable organic dye laser. *Nature (London)* **240,** 483.

18. Berns, M. W. (1975). Biological microirradiation: Classical and laser sources. *In* "Lasers in Physical Chemistry and Biophysics" (J. Joussot-Dubien, ed.), pp. 389–401. Elsevier, Amsterdam.

19. Berns, M. W., Aist, J., Edwards, J., Strahs, K., Girton, J., McNeil, P., *et al.* (1981). Laser microsurgery in cell and developmental biology. *Science* **213,** 505–513.

20. Monajembashi, S., Cremer, C., Cremer, T., Wolfrum, J., and Greulich, K. O. (1986). Microdissection of human chromosomes by a laser microbeam. *Exp. Cell Res.* **167,** 262–265.

21. Ponelies, N., Bautz, E. K. F., Monajembashi, S., Wolfrum, J., and

Greulich, K. O. (1989). Telomeric sequences derived from laser-microdissected polytene chromosomes. *Chromosoma* **98**, 351–357.

22. Hadano, S., Watanabe, M., Yokoi, H., Kogi, M., Kondo, I., Tsuchiya, H., Kanazawa, I., Wasaka, K., and Ikeda, I.-E. (1991). Laser microdissection and single unique primer PCR allow generation of regional chromosome DNA clones from a single human chromosome. *Genomics* **11**, 364–373.

23. Djabali, M., Nguyen, C., Biunno, I., Oostra, B. A., Mattei, M.-G., Ikeda, J.-E., and Jordan, B. R. (1991). Laser microdissection of the fragile-X region: Identification of cosmid clones and of conserved sequences in this region. *Genomics* **10**, 1053–1060.

24. Lengauer, C., Eckelt, A., Weith, A., Endlich, N., Ponelies, N., Lichter, P., Greulich, K. O., and Cremer, T. (1991). Painting of defined chromosomal regions by *in situ* suppression hybridization of libraries from laser-microdissected chromosomes. *Cytogenet. Cell Genet.* **56**, 27–30.

25. Letokhov, V. S. (1985). Laser biology and medicine. *Nature* **316**, 325.

26. Gorodetsky, G., Kazyaka, T. G., Melcher, R. L., and Srinivasan, R. (1985). Calorimetric and acoustic study of ultraviolet light laser ablation of polymers. *Appl. Phys. Lett.* **46**, 828.

27. Pirotta, V., Jäckle, H., and Edström, J. E. (1983). Microcloning of microdissected chromosome fragments. *In* "Genetic Engineering: Principles and Methods" (J. K. Setlow and A. Hollaender, eds.), Vol. V., pp. 1–7. Plenum, New York.

Chapter 4

Molecular Cloning of Microdissected Chromosomal DNA

Introduction

Efforts to develop recombinant DNA methods to clone specific genes directly from microdissected eukaryotic chromosomes began over a decade ago. These efforts have been enhanced by the rapid advances in PCR technology. The polymerase chain reaction produces DNA in quantities sufficient for genetic and molecular analysis from minute quantities of template DNA. The use of microdissection methods to prepare chromosomal DNA for cloning has facilitated the characterization of several novel genes that have been implicated in inheritable human diseases. In this regard, microdissection and cloning offer the potential for identifying genes that have been localized precisely on a chromosomal map but for which DNA sequence information and specific molecular probes are not available.

In the previous chapter, we described three different methods that have been used to obtain DNA from chromosomes using microdissection. In this chapter, we describe the procedures that have been employed to generate recombinant DNA clones from a few chromosomal fragments. To date, three distinct methods have been described for generating chromosome libraries from microdissected chromosomal DNA. In the first method, DNA is extracted from chromosome fragments in a nanoliter aqueous drop, digested with a suitable restriction enzyme, and cloned directly into a λ phage vector (microcloning). The second method is an extension of the first that utilizes PCR. Restriction enzyme digested, microdissected DNA is cloned into a plasmic vector, the inserts are amplified by PCR using vector-specific primers. The third method involves direct PCR amplification of DNA contained in dissected fragments, using carefully designed oligonucleotide primer sequences that occur frequently in the genome being studied. The PCR product is used to probe a more complete recombinant DNA library. The efficiency of cloning varies among these methods, but the end result is the construction of a library of clones specific to a chromosomal region. The cloned DNA sequences can be analyzed directly or used to probe more comprehensive recombinant libraries [genomic, chromosome specific, yeast artificial chromosomes (YAC), or cDNA]. Potential genes of interest are characterized by a series of Southern blot hybridizations and direct DNA sequencing, and localized on a chromosomal map by *in situ* hybridization.

Since the focus of this manual is microdissection and microcloning, discussion of specialized techniques used in conjunction with each method is kept to a minimum. For a detailed historical and theoretical background of these procedures, other sources, such as the manuals by Sambrook *et al.* (1), Davis *et al.* (2), and Ausubel *et al.* (3), are highly recommended. In these protocols, we emphasize PCR, because this procedure facilitates cloning of minute quantities of DNA, and fluorescent *in situ* hybridization, because this method allows rapid characterization of recombinant clones derived from specific chromosome regions.

CLONING DNA FROM MICRODISSECTED CHROMOSOMAL DNA FRAGMENTS

In each of the methods described here, the dissected chromosomal DNA fragments are collected in a drop containing a few nanoliters of an aqueous buffer solution. Then the chromosomal DNA in the microdrop is extracted to remove nucleoproteins partially or completely and to prepare DNA for subsequent cloning.

Method 1. Direct Cloning of DNA from Microdissected Chromosomal Fragments

In this first method (see Figure 4.1), microdissected chromosomal DNA is extracted with proteinase K and subsequently digested with a suitable restriction enzyme such as *Eco*RI. The smaller fragments generated (estimated to be 3–4 kb) are ligated to a λ phage vector molecule, a reaction that is not dependent on large quantities of insert DNA and, therefore, is applicable to microcloning. The recombinant phage permits easy propagation of the inserts via appropriate host bacteria. This method was used effectively in the earlier microdissection experiments to clone DNA from polytene chromosomes of *Drosophila melanogaster* (4), as well as from mouse (5, 6) and human (7, 8) metaphase chromosomes. This method requires a large number of fragments (50–200) to provide sufficient starting material for cloning, particularly if nonpolytenized chromosome bands are dissected. The number of clones obtained by this method is relatively small and represents a small percentage of the DNA sequences contained in the dissected chromosomal fragment.

Reagents and Materials

Agarose (e.g., FMC, Boehringer-Mannheim)
Aqueous buffer—4 volumes 87% glycerol, 1 volume 0.05 M sodium potassium phosphate buffer (pH 6.8)
Bacterial and phage growth media (Difco)
Bacterial plates (Fisher Scientific)
Buffer-saturated distilled phenol—0.1 M Tris (pH 8.0), 0.2% 2-mercaptoethanol, 0.1% hydroxyquinoline
Chloroform—24 : 1 (v/v) mixture of chloroform and isoamyl alcohol
Cloning vector—phage λ gt10 or λ MN641 (Stratagene)
Coverslips, sterile, clean

Chapter 4

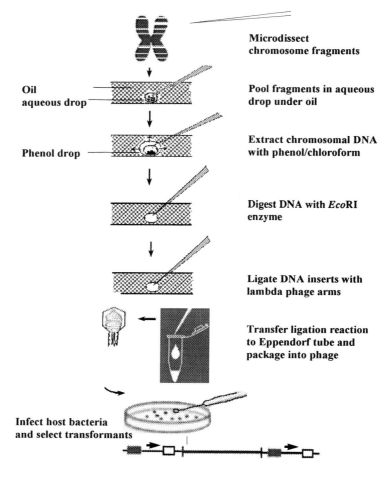

Figure 4.1 Strategy for directly cloning microdissected chromosomal DNA.

Extraction buffer—500 μg/ml proteinase K, 10 mM Tris (pH 7.5), 10 mM NaCl, 0.05% SDS
Gel electrophoresis equipment (e.g., Bio-Rad)
Host bacteria *E. coli* hflA⁻ (e.g., C600, LE392)

Incubation chambers, 14°–16°C, 37°C, 70°C (humidified)
Microcentrifuge tubes, 0.5 ml
Microdissection apparatus (see Chapter 3 for details)
Phage packaging extract (Stratagene)
Restriction enzymes (e.g., *Eco*RI, *Hin*dIII) and buffers (New England Biolabs)
SM buffer—10 mM Tris–HCl (pH 7.0), 10 mM MgSO$_4$, 0.01% gelatin
T4 DNA ligase and buffer (New England Biolabs)

Protocol 1.1. Direct Cloning into λ Phage (4, 5)

Preparation of Chromosomal Fragments and Microdissection

1. Prepare chromosome spreads from a suitable cell source on sterile clean coverslips (see Chapter 2).

2. Prepare microdissection and volumetric (liquid transfer) micropipettes (see Chapter 3). The dissection pipette tip should be <0.5 μm, whereas the tip diameter of the volumetric pipette is 1–2 μm. Both dissection and volumetric pipettes are filled with oil. The aqueous solutions are drawn from the tip under oil using suction generated with a microsyringe.

3. Use the method of preference to microdissect chromosomal fragments (10–20 polytene chromosomal fragments are sufficient, whereas at least 100 fragments are needed to clone from mammalian chromosomes). Fragments should be collected and extracted as described in the next section. Cutting and collecting 10–15 fragments in a microdrop takes approximately 1 hour, including the time spent to search for, find, and position the desired chromosome in the appropriate orientation for cutting.

DNA Extraction and Cloning

1. Using a volumetric micropipette, introduce 1–2 nl aqueous buffer and 6 nl extraction buffer under 200 μl paraffin oil that has

been placed on a concave depression in a glass slide. The slide fits into the rotating stage (Chapter 3). Silicone oil works as well as paraffin oil. If the oil-chamber method is used (Oil Chamber Method, Chapter 3) the chamber is prefilled with oil first.

2. Dissect the fragment, secure to the pipette tip with suction, transfer to the aqueous drop, and pool 5–20 fragments in one drop. Allow to rehydrate for 10–15 minutes.

3. Use a liquid transfer pipette to transfer the fragments to the extraction droplet. Alternatively, transfer one volume (7 nl) of proteinase K/SDS solution to the DNA drop. Mix solutions mechanically using the pipette.

4. Allow fragments to incubate for at least 15 minutes in the extraction buffer at room temperature. If the fragments are still visible in the drop, this step may be extended to 90–120 minutes at 37°C in a moistened chamber.

5. Using a second pipette, introduce 10–15 nl phenol adjacent to the extraction drop. Mix the drops and incubate briefly. The phenol phase surrounds the aqueous drop. Remove the phenol after a few seconds because it dissolves slowly into the oil phase. Repeat this step two more times with fresh phenol.

6. Remove the residual phenol by applying a drop of chloroform on top of the aqueous drop. Separate the chloroform phase using a micropipette.

7. Transfer the aqueous DNA drop to a clean fresh slide and place under oil.

8. Introduce a drop of enzyme buffer containing 100 U/μl *Eco*RI restriction endonuclease and place next to the DNA drop.

9. Using the micropipette, transfer a portion of *Eco*RI enzyme solution from the supply drop to the DNA drop. The ratio of enzyme volume to DNA volume should be 1:5. Incubate for 90 minutes at 37°C in a moistened chamber.

10. Inactivate the restriction enzyme by transferring the slide to a 70°C chamber for 20 minutes. The enzyme also may be inactivated by phenol extraction.

11. Transfer a 4-nl drop containing vector arms (200 μl/ml *Eco*RI-digested λ phage) and 4 nl T4 DNA ligase (2 mM ATP, 1 U/μl T4 ligase) to the slide and mix with the DNA drop.

12. Incubate under oil at 14–16°C for 12–18 hours.

Propagation and Amplification of Cloned DNA Inserts

1. Remove the ligation reaction from oil with a micropipette and place in 2–3 μl buffer containing *in vitro* phage packaging extract in a 0.5-ml microcentrifuge tube. Incubate this mixture for 2 hours at 22°C.

2. Dilute the packaging reaction and use aliquots to propagate and determine the titer of the library as described elsewhere (1, 2). Briefly,

   a. Add 1 μl packaging reaction to 99 μl SM buffer and 200 μl hflA$^-$ host bacteria (e.g., LE392) and incubate at room temperature for 20 minutes.

   b. Add 2.5 ml sterile LB top agar (at 45°C) supplemented with 10 mM MgSO$_4$. Pour onto LB agar plates.

   c. Incubate plates at 37°C for 12–16 hours. Plaques will be visible the next day.

3. Count the number of plaques of each dilution and compare to control plates (phage without inserts) to calculate the size of the recombinant library. (See the section on microclone analysis and characterization.)

Troubleshooting and Solutions

A summary of the most frequently encountered problems when performing microdissection and cloning experiments is provided here.

1. *Presence of clear plaques that do not contain inserts (high background).* This problem is caused primarily by self-religation of phage DNA vector or the presence of spontaneous λ phage mutants, and can be minimized by

    increasing the ratio of cloning vector to insert
    using highly purified phage vector (molecular biology grade)
    reducing the initial ligation reaction to a few nanoliters volume; when the concentration of reactants is increased, a better insert–vector ligation is achieved

2. *The number of recombinant DNA clones obtained is very small and does not represent the entire genomic sequence contained in the dissected fragment.* This problem could be caused by

    excessive acid fixation of chromosomes during preparation of spreads
    degradation of chromosomal DNA on the slides through aging in air or ethanol
    insufficient starting chromosomal DNA
    incomplete ligation or low cloning efficiency

    These problems can be overcome by taking the following precautions:

    prepare chromosomes immediately before microdissection
    prepare chromosomes using conditions that ensure that the DNA is intact and of the highest quality
    make sure that all slides containing chromosome spreads, handling pipettes, and solutions are cleaned thoroughly and sterile, presence of contaminating microorganisms may result in DNA degradation and clones that contain foreign DNA.
    increase initial number of chromosomal fragments dissected (usually about 100 fragments are sufficient
    use the calculations discussed on pages 107–109 to determine the time, reaction volume, and stoichiometry of reactants (DNA inserts, ligase, and vector) in the ligation reaction carefully to achieve optimal cloning of all inserts present

3. *The average size of cloned DNA is less than that predicted.* Contributing factors are

   DNA degradation

   excessive fixation and nucleic acid depurination during preparation of chromosomes (as discussed earlier, these problems are avoidable)

   the recognition sequence of the restriction enzyme used to digest the fragment DNA occurs at higher frequency than is estimated; select another restriction enzyme to prepare the inserts provided that the vector contains a compatible cloning site

Method 2. Ligation of Microdissected Chromosomal DNA with Plasmid Vector or Linker–Adaptor and PCR Amplification (9–11)

In this method (see Figure 4.2), the microdissected DNA is digested with a restriction endonuclease that cuts frequently within the genome (e.g., *Rsa*I, *Hpa*II). The fragments generated are ligated into a linearized plasmid. The DNA then is amplified using PCR with oligonucleotide primers that are complementary to plasmid vector sequences. The amplified DNA inserts are cleaved with a second enzyme (*Eco*RI) that flanks the cloning site. The inserts subsequently are subcloned into a second plasmid vector (e.g., pUC plasmids) to generate the library. This procedure was employed first to clone specific genes from human chromosome 8 from Langer–Giedion syndrome patients (9–11). The procedure offers several advantages. Cloning in plasmid vectors is easier than cloning in phage and clones can be characterized more rapidly. Also, because of the smaller size of the vector, DNA inserts in plasmid vectors are more amenable to enzymatic amplification than are inserts in phage vectors. The introduction of PCR allows the amplification of a population of clones that otherwise would go undetected. Finally, the use of PCR significantly reduces the number of chromosomal fragments initially required for cloning. Another method that has been used employs ligation of dissected DNA to linker–adaptors rather than to plasmid before PCR amplification (12).

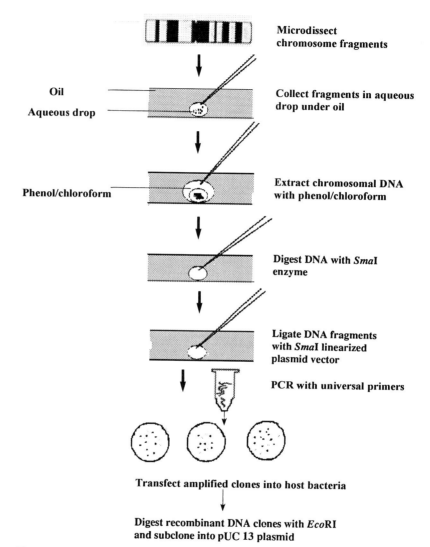

Figure 4.2 Strategy for cloning microdissected chromosomal DNA by ligation with a plasmid vector and PCR amplification.

Reagents and Materials

Agarose (FMC)
Aqueous buffer (B)—20 mM Tris–HCl (pH 8.0), 10 mM $MgCl_2$, 50 mM NaCl
Bacterial and phage growth media and plates (Difco)
BamHI linearized Bluescribe® plasmid (BRL)
Buffer-saturated phenol and chloroform—0.1 M Tris–HCl (pH 8.0)
10× CIP buffer— 0.5 M Tris–Cl (pH 9.0), 10 mM $MgCl_2$, 1 mM $ZnCl_2$, 10 mM spermidine
Cloning vectors (e.g., plasmid pUC 13, 18) (Pharmacia), linearized with SmaI
Containment hood
Dithiothreitol
DNA extraction buffer (A)—10 mM Tris–HCl (pH 7.5), 10 mM NaCl, 0.1% SDS, 1% glycerol, 500 µg/ml proteinase K
Gel electrophoresis equipment (Bio-Rad)
Host bacteria (e.g., DH5 α) (BRL)
Humidified (moistened) chamber, for example, Petri dish lined with water-saturated tissue (e.g., Kimwipes) or incubator
Isopropanol
LB medium
MboI DNA linker–adaptor (Boehringer Mannheim)
Microcentrifuge tubes, 0.5 ml
Microdissection apparatus (as discussed in Chapter 3)
Nitrocellulose or biomembrane filters (Fisher Scientific)
Oligonucleotide primers (M13 primers, MboI linker primers) (Boehringer-Mannheim)
PCR reaction kit (e.g., AmpliTaq; Perkin Elmer/Cetus)
Reagents for Southern blot hybridization—6× SSC, 0.1 M EDTA, 5× Denhardt's solution, 0.5% SDS, 100 µg/ml denatured salmon sperm DNA
Restriction enzymes (e.g., RsaI, SmaI, EcoRI, MboI) and enzyme buffers
Sephacryl S200 columns (Pharmacia)
Sephadex G-50 column (Pharmacia)

T4 DNA ligase and T4 ligase reaction mix—250 mM Tris–HCl (pH 7.6) 50 mM MgCl$_2$, 5 mM ATP, 5 mM dithiothreitol, 25% polyethylene glycol 8000 (New England Biolabs)
columns (Pharmacia)
Taq DNA polymerase (e.g., U.S. Biochemicals)
TE buffer—10 mM Tris–HCl (pH 7.0), 0.5 mM sodium EDTA
Thermal cycler (Perkin Elmer/Cetus; Ericom, Inc.)
Water bath, 42°C

Protocol 2.1. Ligation of Microdissected DNA with Plasmid Vector, PCR Amplification, and Cloning (9–11)

In this protocol, microdissected DNA first is cloned into a plasmid vector and then is amplified enzymatically using universal primer binding sites in the vector that flank the insert DNA. The genomic DNA is digested with an enzyme, for example, *Rsa*I that generates fragments of an average length of 1000–2000 bp, an optimal size for PCR amplification. Also, the *Rsa*I site occurs infrequently in repeat sequence DNA, thus minimizing preferential cloning of repeat sequences. The plasmid vector (e.g., pUC 18) contains M13 primer sequences that flank the multiple cloning site. Thus, only one pair of oligonucleotide primers is used for the simultaneous amplification of many different DNA sequences.

Preparation of Chromosomes and Microdissection

1. Prepare chromosome spreads from a suitable cell source on a sterile clean coverslip (see Chapter 2).

2. Prepare sterile clean microdissection (0.5-μm tip diameter) and liquid transfer (1- to 2-μm tip diameter) micropipettes (see Chapter 3).

3. Introduce DNA collection (buffer A) and extraction solutions on the dissection slide and place the slide under a moistened chamber to prevent evaporation.

4. Collect a sufficient number of chromosomal fragments for cloning. Approximately 40–50 mammalian chromosomal fragments are required. If polytene chromosomes are used, 1–10 fragments are sufficient.

DNA Extraction and Cloning

1. Pool the dissected fragments directly into a 1- to 3-nl microdrop of extraction buffer A that is situated under a moistened chamber on the dissection slide

2. Incubate for at least 30 minutes in the extraction buffer or until the fragments are no longer visible.

3. Carefully extract the DNA drop three times with 2–3 volumes of saturated phenol (see Reagents and Materials under Method 1). Avoid excessive pipetting that results in shearing of the DNA.

4. Extract residual phenol with chloroform.

5. Transfer restriction enzyme (RsaI, final concentration 6 U/μl) from a supply drop to the DNA drop. Incubate slide in a moistened chamber at 37°C for 2–3 hours. The restriction enzyme may be replenished and incubation continued for another 2 hours.

6. Inactivate the enzyme either by heat denaturation at 65°C for 3 minutes or by phenol extraction.

7. Add fivefold molar excess SmaI-linearized pUC 18 vector to the digested fragments.

8. For ligation, add 0.08 U T4 ligase and ligase mix. Incubate mixture at 15°C overnight.

Polymerase Chain Reaction Amplification

In this procedure and all others that involve PCR, all PCR reactions must be performed under a containment hood or in a dust-free area.

1. Combine the ligation reaction mixture with 2 µl water and transfer to a 0.5-ml microcentrifuge tube.

2. Add 1 µmol each of the M13 universal forward and reverse primers. The following M13/pUC forward and reverse sequencing primers are commonly used:

   forward sequencing primer: 5'-ACTGGCCGTCGTTTTAC-3'

   reverse sequencing primer: 3'-GTCCTTTGTCGATACTG-5'

3. Adjust the sample volume to make the total reaction volume 50 µl; use PCR buffer containing 20 mM Tris–HCl (pH 7.8), 5 mM $MgCl_2$. Add dithiothreitol to 1 mM and 2.5 mM of each of the four deoxynucleoside triphosphates (PCR reaction kit).

4. Add 2.5 U (0.5 µl) *Taq* DNA polymerase enzyme. Perform PCR amplification of DNA inserts using the following thermal profile:

   a. Denature DNA templates at 94°C for 5 minutes in the initial cycle (1 minute in the remaining cycles).

   b. Anneal primers to templates at 55°C for 1 minute.

   c. Perform primer extension at 72°C for 2 minutes.

   d. Repeat Steps a–c for 25–30 cycles.

5. Remove amplified products and purify DNA by phenol extraction (mix well with an equal volume of water-saturated phenol, centrifuge at 3000 rpm to separate the aqueous phase), followed by ethanol precipitation (mix aqueous phase with 2–3 volumes cold absolute ethanol, freeze at −70°C for 10 minutes, pellet DNA by centrifugation at 10,000 rpm for 20 minutes) (1, 3). Resuspend pellet in *Eco*RI reaction buffer and add 20 U *Eco*RI enzyme to digest at 37°C for 2 hours in a final volume of 100 µl. Analyze an aliquot of digested DNA by agarose gel electrophoresis to check size range of DNA inserts.

6. Further purify the cloned DNA inserts by spinning through two

successive Sephacryl S200 columns (Pharmacia) equilibrated with ligation buffer (eluate volume is 100–150 µl).

7. Subclone *Eco*RI fragments in pUC 13 plasmid vector. Ligation is performed by mixing 20 ng dephosphorylated *Eco*RI linearized vector and 17 µl eluate containing DNA inserts. Add 0.08 U T4 ligase and incubate at 15°C overnight (1–3).

8. Propagate recombinant plasmid in appropriate host bacteria (e.g., DH5 α) after transfection. The following transfection protocol can be used:

   a. Prepare competent bacterial cells as described elsewhere (1–3).

   b. Mix 100 ng plasmid vector DNA with 200 µl competent bacteria (DH5 α) in a sterile Eppendorf tube.

   c. Incubate on ice for 30 minutes.

   d. Heat shock bacterial cells by transferring to a 42°C water bath for 2 minutes.

   e. Add 1 ml LB medium and grow for 45 minutes at 37°C with shaking.

   f. Plate 100 µl on agar plates containing 100 µg/ml ampicillin and 50 µg/ml 5-bromo-4-chloro-3-indolyl-β-galactopyranoside (XGal) and incubate overnight at 37°C.

   g. Isolate DNA from single colonies for detailed clone analysis (see Protocols 4.1, 4.2, and 5.2).

Protocol 2.2. Ligation of Microdissected DNA with Linker–Adaptor, PCR Amplification, and Cloning (12)

A linker–adaptor commonly used in this method is a 24-nucleotide molecule with a terminus complementary to the *Mbo*I enzyme recognition sequence. The *Mbo*I linker–adaptors have the sequence

5′-GATCTGTACTGCACCAGCAAATCC-3′

3′-ACATGACGTGGTCGTTTAGG-5′

The linker–adaptor is formed by annealing a mixture containing 100 µg/ml of each oligonucleotide at 58°C for 1 hour. Prior to the annealing step, the 24-mer oligonucleotide component carrying the 5′ *Mbo*I complementary overhanging end is phosphorylated. This step is necessary to prevent the formation of tandem arrays of linkers during ligation, which would interfere with the subsequent amplification of chromosomal DNA (12).

Ligation of Microdissected DNA with Linker Primers

1. Perform Steps 1–4 of Protocol 2.1.
2. Place chromosome fragments into 1–2 nl aqueous buffer B.
3. Add SDS and proteinase K to final concentrations of 0.1% and 500 µl/ml, respectively, from a supply drop that is placed next to the aqueous drop. Incubate for 1 hour at 37°C.
4. Extract the drop containing the DNA three times with phenol saturated with the aqueous buffer B.
5. Remove any residual phenol by extraction with 1 volume of chloroform and transfer to a new oil drop.
6. Using a clean micropipette, introduce 0.1 volume of *Mbo*I enzyme (40 U/µl).
7. Incubate the slide in a moistened chamber at 37°C for at least 2 hours.
8. Return slide to the microscope and add 0.25 volume of T4 ligase mix.
9. Add 0.25 volume of 9.5 U/µl T4 DNA ligase and an equal volume of the *Mbo*I linker–adaptor (100 µg/ml in TE buffer).
10. Allow ligation reaction to incubate at 19°C for 16 hours.

PCR Amplification

1. Transfer 3 µl TE buffer to the slide and mix well with the microdrop containing the ligation reaction products.

2. Collect the drop with an Eppendorf pipette and recover into a 0.5-ml microcentrifuge tube.

3. Adjust volume to 50 µl with PCR buffer containing 2.5 U *Taq* DNA polymerase and 2 µmol of the 20-mer component of *Mbo*I linker–adaptor as the primer.

4. Carry out 35 cycles of PCR amplification as follows:

    a. Denature at 94°C for 1 minute.

    b. Anneal primer to templates at 52°C for 2 minutes.

    c. Perform extension at 72°C for 3 minutes.

5. Extract amplified DNA with phenol followed by ethanol precipitation, as described earlier in Step 5 (see p. 90).

6. Remove amplified chromosomal DNA by digestion with *Mbo*I restriction enzyme (use enzyme and buffer according to the supplier's instructions).

7. Dissolve the DNA in a minimum volume of 10 mM Tris–HCl (pH 8.0). Incubate DNA at 37°C for 30 minutes in the presence of 0.01 U calf intestine alkaline phosphatase (CIP) and 1× CIP buffer to remove phosphates from 5' ends of DNA. Extract twice with phenol/chloroform. Pass the aqueous phase through a spin column of Sephadex G-50 equilibrated in TE buffer (1–3). Precipitate the DNA with ethanol. The PCR product is now ready for ligation.

8. Ligate dephosphorylated inserts to *Bam*HI linearized Bluescribe plasmid (BRL) and transfect competent bacteria as described in Protocol 2.1 in this chapter.

9. Proceed to identify clones containing unique chromosomal sequences as described on pp. 105–106 of this chapter.

Troubleshooting and Solutions

Most of the problems that are described for Method 1 also can occur using Method 2. In addition, some problems are particularly pertinent to PCR methodology:

1. *Low number of recombinant clones:*

    optimize the concentrations of DNA, ligase enzyme, and linearized vector in the ligation reaction step
    increase the initial number of PCR cycles; running an extra round (35 cycles) of amplification usually results in better detection of clones as well as amplification of small populations of clones that were not detected in the first round amplification product
    test cloning efficiency using control DNA ligation

2. *The presence of a large number of very small fragments (less than 50 bp)*. This problem presumably is caused by:

    occurrence of *Rsa*I restriction sites at higher frequency within the fragment chosen for dissection
    degradation of DNA
    preference of PCR for amplification of shorter DNA sequences

    The choice of another enzyme, for example, *Hpa*II, has been shown to improve the number of inserts obtained by this method (13). Optimizing PCR annealing and extension temperatures and times may be helpful.

Method 3. PCR Amplification of Microdissected Chromosomal DNA Fragments Followed by Probing a Complete Recombinant Library (14–16)

The direct approach of this method (see Figure 4.3) permits the isolation of DNA sequences from a specific chromosomal region by amplification of the microdissected DNA with PCR using a set of universal primers that contain a series of degenerate base sequences. The amplification product should represent a heterogeneous population of DNA sequences. This PCR product can be labeled by means of the random priming method (17) and used as a probe to screen a complete phage or cosmid library. Direct PCR amplification of microdissected DNA has been employed to isolate genes from band 2F of the *Drosophila* polytene X chromosome using degenerate "universal" primers (14). Also, libraries of specific human chromosome bands were constructed using universal primers and PCR (15). The use of *Alu*-specific primers to screen for human genes in complex DNA mixtures has been described previously (18, 21). Probes generated from microdissected chromosomal DNA fragments with *Alu* PCR have been used to localize specific chromosomal sequences in YAC recombinant clones (22, 23). These probes also were used to generate sequence tagged sites (STSs) from discrete chromosomal regions (24). Direct PCR amplification of microdissected DNA has the following advantages over the methods described earlier. By directly amplifying picogram amounts of DNA, time-consuming steps that involve restriction enzyme digestion, ligation, and subcloning are obviated. This approach minimizes the loss of minute starting quantities of DNA during the various manipulations employed in the other techniques. Most importantly, the clones constituting the final library contain DNA inserts of large size. However, these primers will amplify minute quantities of contaminant DNA so special precautions are necessary (see Table 4.1).

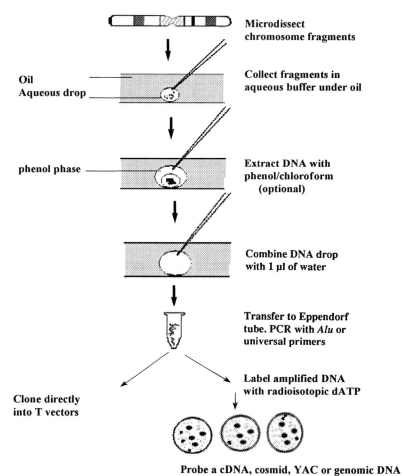

Figure 4.3 Strategy for cloning microdissected chromosomal DNA using PCR amplification, and subsequently probing a complete recombinant library.

Table 4.1 Precautions to Be Taken When Amplifying Chromosomal Fragment DNA Using "Universal" Primers

1. Chromosome spreads should be dropped on coverslips that are acid washed, then washed extensively with sterile water.
2. The oil that fills the microdissection system should be filter sterilized.
3. Glass cutting needles should be placed under UV light overnight.
4. All solutions should be filter sterilized and placed under UV light for several hours.
5. Designated pipettes should be used only for chromosome fragment amplification; pipette tips and reaction tubes should be sterile.
6. Test each lot of *Taq* DNA polymerase in a PCR reaction that does not contain template DNA. After three rounds of amplification (see Protocol 3.2), no PCR product should be detectable. Any PCR product seen in an agarose/ethidium bromide gel may indicate that the *Taq* polymerase reagents contain contaminating bacterial DNA. Another lot should be tested.
7. A "water control" (reaction without DNA template) PCR should always be performed in tandem with the test reaction to ensure that the test reaction PCR DNA product was generated from chromosomal fragments.

Reagents and Materials

Bacterial and phage growth media (Difco)
Bacterial plates (Fisher Scientific)
Containment hood
DNA extraction buffer—10 mM Tris–HCl (pH 7.5), 10 mM NaCl, 0.1% SDS, 1% glycerol, 500 μg/ml proteinase K (optimal)
Genomic or individual chromosome phage library, complete (a cDNA library may be used also) (Stratagene, American Type Culture Collection)
Glass micropipettes, slides, coverglass, and pipette tips, sterile
Host bacteria for phage growth
Microdissection apparatus (see Chapter 3)
Paraffin oil, sterile, filtered
PCR reaction kit (e.g., GeneAmp from Perkin Elmer/Cetus)
Primers (universal, *Alu* primers) (GenoSys Corp.)

Reagents for agarose gel electrophoresis
Reagents for Southern blot hybridization
Taq (DNA) polymerase and *Taq* polymerase 10× buffer—500 mM KCl, 100 mM Tris–HCl (pH 8.3), 20 mM MgCl$_2$, 0.1% gelatin
Thermal cycler (e.g., Perkin Elmer/Cetus; Ericom, Inc.)
Water, sterile, distilled, deionized

Protocol 3.1. Preparation of Chromosomal DNA for Amplification

1. Prepare chromosome spreads of the highest quality from suitable cells, for example, peripheral blood lymphocytes (see Chapter 2).

2. Prepare microdissection needles (0.5-μm tip diameter) and liquid transfer micropipettes (1- to 2-μm tip diameter). Fill both pipettes with oil from the back end using 26-gauge spinal needles.

3. Using an Eppendorf pipette, introduce 0.5–1 μl supply drops of extraction buffer under oil that is placed on the depression slide. The aqueous drop must touch the surface of the slide. The depression slide is placed next to the dissection slide on a circular rotating stage. Bring the tips of the transfer micropipettes under the oil and fill with aqueous solutions using microsyringes and suction. Transfer nanoliter drops for collection and extraction of DNA.

4. Microdissect a sufficient number of chromosomal fragments (40–80 human metaphase chromosome fragments) and pool into the aqueous drop under oil.

Frequently we use a quick alternative to collect the dissected fragments directly into the PCR reaction tube. This short cut avoids the manipulation of microdrops, which is time consuming. Briefly, while the dissected fragment is still adherent to the tip of the micropipette, the tip is broken into the bottom of a microcentrifuge tube containing 20 μl sterile deionized water. After the DNA is rehydrated and dissolved (several hours to overnight), the broken glass is pelleted by centrifugation. The aqueous solution is transferred to a clean tube.

5. Use fragments in PCR amplification reaction immediately or freeze at −70°C.

Protocol 3.2. PCR Amplification of Microdissected Chromosomal DNA Using "Universal" Primers

Universal primers comprise a mixture of oligonucleotide sequences that lack absolute complementarity to the target template sequences. The mixture contains multiple degenerate bases. When using these primers in PCR amplification and selecting lower PCR annealing temperatures (46–48°C), complementary sequences that contain significant homologies can be amplified. For example, the following primer was used to generate a probe library from *Drosophila* polytene chromosome fragment (14):

5'-TTGCGGCCGCATTNNNNTTC-3'

With complete degeneracy at positions 4, 5, 6, and 7 from the 3' end, authentic DNA sequences from complex mixtures were isolated. A similar primer also has been described in PCR amplification of microdissected human chromosomes (16, 25). This 22-mer, however, contains complete degeneracy at six nucleotides (positions 11–16) with the following sequence:

5'-CCGACTCGAGNNNNNNATGTGG-3'

In general, a nonspecific primer should have the following structure from 5' to 3' ends: two or more protective (cap) nucleotides (e.g., TT and CCGA in the primers shown), followed by a restriction endonuclease recognition site (*Not*I and *Xho*I, respectively, in the primers shown), 4–6 nucleotides with complete degeneracy, and, finally, a terminal 3' triplet sequence that occurs at high frequency in the genome of the species from which the DNA is amplified.

1. To the extracted DNA (Step 5, Protocol 3.1), add 10 μl 10X *Taq* DNA polymerase buffer. Adjust volume to 50 μl with sterile water.

2. Add 60 ng primers (primer concentration may be increased up to 1 µg, which appears to increase amplification during the earlier cycles).

3. Boil mixture for 5 minutes and rapidly chill on ice for 3 minutes to form initial primer–template complex.

4. Add 8 µl each of the four 1.25 mM deoxynucleotides (final concentration is 0.1 mM each).

5. Add 0.5 µl 5.0 U/µl *Taq* DNA polymerase.

6. Adjust the final volume to 100 µl with sterile deionized water.

7. Overlay with 100 µl paraffin oil.

8. Run thermal cycler with the following temperature cycle profile:

   a. The first 10 cycles, denature at 95°C for 1 minute, anneal at 27°C for 1 second, and extend primers for 2 minutes at 67°C.

   b. The next 35 cycles, denature at 95°C for 1 minute, anneal at 50°C for 1 minute, and extend primers at 72°C for 2 minutes.

9. Remove 10 µl of the amplified products to use as template for a second PCR reaction, as described by repeating Steps 1–7 and running a second round of amplification identical to Step 8b.

10. If necessary, transfer 5 µl of the second round of PCR reaction mixture and subject it to a third and final round of amplification. This step is performed if the amplification products are difficult to visualize on agarose gel after electrophoresis.

    Note: If the PCR reaction in Steps 9 and 10 does not yield a good amplification product, the entire PCR reaction product in Step 8 may be used for subsequent rounds of amplification following its purification on a Centricon spin column.

11. The PCR product is then labeled with ^{32}P-labeled nucleoside triphosphate by the random priming method (17) and used to probe a more comprehensive genomic or appropriate cDNA library.

Protocol 3.3. PCR Amplification Using Human Alu Sequence-Based Primers (18)

The *Alu* repeats (SINES) and L1 repeats (LINES) are short, interspersed, repeated DNA elements distributed throughout the genomes of primates. *Alu* elements are dimeric in structure and are composed of two tandemly arranged halves. The right half contains an additional 31 bp relative to the left half. *Alu* repeats are approximately 300 bp in length and are present at a copy number in excess of 10^6 per haploid human genome equivalent. L1 repeats, however, are present in 10^4-10^5 copies per genome. *Alu* and L1 repeats occur preferentially in light Giemsa bands and dark Giemsa bands, respectively, as described in Chapter 1. *Alu* and L1 core sequence-based primers were used by Nelson et al. (18) and Ledbetter et al. (20) to amplify human DNA sequences located between the two inverted repeats in hamster–human hybrid cells containing specific human chromosomes. The most commonly used repeat sequence primers utilized for PCR were described by Nelson et al. (18):

Alu-559 5'-AAGTCGCGGCCGCTTGCAGTGAGCCGAGAT-3'
Alu-517 5'-CGACCTCGAGATCT(C/T)(G/A)GCTCACTGCAA-3'
L1Hs   5'-CATGGCACATGTATACATATGTAAC(T/A)AACC-3'

For amplification of human chromosomal DNA with *Alu* primers, we have obtained consistently good results using the following method (15).

1. Add 10 µl 10X *Taq* DNA polymerase buffer to the drop containing chromosomal DNA. Add the following:

   125 ng human DNA-specific *Alu* primers
   1.0 mM each of the four deoxynucleotides (final concentration is 0.1 mM)

   Adjust volume to 100 µl with sterile water.

2. Add 2.5 U *Taq* DNA polymerase.

3. Overlay with 100 µl paraffin oil.

4. Run 35 cycles using the following temperature cycle profile:

   Denature DNA at 94°C for 30 seconds, except for the first cycle (5 minutes denaturation).
   Anneal at 55°C for 75 seconds and extend primers for 3 minutes at 72°C.

5. Withdraw 10 µl of the first round of amplification to use as template for a second PCR, following Steps 1–4 in this protocol.

6. Transfer 10 µl of the second round PCR reaction and perform a third round of amplification.

   Note: If the PCR reaction product in Steps 4 and 5 is still low, use the entire PCR reaction product in Step 3 for subsequent rounds of amplification, following its purification and concentration on a Centricon DNA spin column.

7. Label the PCR product with ^{32}P-nucleoside triphosphate by the random priming method (17) and use to probe a more comprehensive genomic or appropriate cDNA library (1, 2) using suppression hybridization with repeat sequence DNA (26).

Troubleshooting and Solutions

This method does not involve cloning. Hence, its efficiency depends entirely on the completeness of PCR amplification with the primer set chosen. The completeness of the library is dependent on the sequence of primers and the PCR reaction conditions (e.g., annealing temperatures during the early cycles and primer extension times), concentration of dNTPs, the quality of DNA polymerase, and, finally, the number of cycles required to amplify sufficient amounts of DNA inserts containing unique sequences (usually 60–90 cycles). Frequent problems with this method include the following:

1. *Cannot visualize amplified products on agarose gel after PCR.*

   increase amount of DNA (number of fragments)

increasing primer concentration (up to 1 μg, particularly with primers containing degenerate sequences) may help increase amplification products during the early cycles of PCR

decrease annealing temperature of the initial thermal cycles of PCR (temperatures in the range of 27–30°C have been reported)

increase number of thermal cycles; remove an aliquot from first round and repeat PCR

consider cutting DNA with a restriction enzyme to generate a homogeneous population of fragments of reasonable lengths that are amenable to uniform PCR amplification; *Hpa*II enzyme has been used to generate amplifiable sequences from human chromosomes (13)

2. *PCR product is small.*

    avoid DNA degradation
    optimize primer extension time and temperature
    use different sets of primers that hybridize to reasonably spaced sequences

3. *A large number of clones are repeat sequences.* The use of an inter-*Alu* primer certainly will result in amplification of repeat sequences as well as unique sequences. This problem can be minimized (but unfortunately not avoided) by

    using *Alu* sequence-based primer sequences that anneal to templates immediately at the junction between repeat and unique sequences [e.g. primer 517 (18)]
    raising annealing temperatures in the initial PCR cycles to ensure the amplification of unique sequences over repeat sequences
    using suppression hybridization of the probe to Cot1 or genomic DNA, which prevents the probe from detecting clones that contain repeat sequences in subsequent hybridization experiments designed to characterize cloned inserts
    using different primer sets

ANALYSIS OF RECOMBINANT CLONES DERIVED FROM MICRODISSECTED CHROMOSOMAL DNA

Once the recombinant DNA library is obtained it should be tested for completeness, complexity, and verification that the cloned DNA came from the microdissected fragment. The following steps provide a guideline for the characterization of a library generated by these methods:

1. Determination of DNA insert size range.
2. Determination of the percentage of recombinant clones containing repeat and unique sequences.
3. Calculation of the percentage of total microdissected DNA cloned.
4. Determination of potential structural gene sequences.
5. Localization of recombinant clones using *in situ* hybridization.

Determination of DNA Insert Size Range

Recombinant DNA is obtained using either the plaque lysate method for phages or a plasmid preparation. [For details of these procedures see Refs. (1–3)]. A minimum of 30–50 clones should be analyzed to determine the quality of the library.

1. Release the insert DNA from the cloning vector by digestion with the appropriate restriction enzyme.
2. Determine the size of cloned inserts by fractionation on an agarose gel using appropriate DNA markers. The size of the cloned insert may be confirmed using a second restriction enzyme site that flanks the cloning site. DNA is visualized on the gel by ethidium bromide staining. If more accurate estimation of cloned fragment size is desired, the fragments are end-labeled with [^{32}P]dATP and run

on polyacraylamide gels. This step becomes necessary when a significant number of small size inserts is obtained ($\leq$ 100 bp).

Estimation of the size range and average size of cloned inserts and determination of the number of independent clones are used to determine the completeness of the library. The restriction enzyme digestion pattern of each clone also gives an estimate of the number of redundant clones in the library. The size range of cloned DNA allows comparison with the DNA size predicted for the restriction enzyme used (e.g., *Eco*RI digestion of genomic DNA yields an average size of 2.9 kb). A smaller insert size indicates that the original chromosomal DNA was degraded or that preferential cloning or amplification of smaller DNA molecules occurred.

Determination of the Percentage of Recombinant Clones Containing Repeat and Unique Sequences

To identify clones with repeat sequences, DNA inserts from 30 to 50 randomly picked clones are transferred to a nitrocellulose filter and probed with total genomic human DNA that is labeled radioisotopically by nick translation (1–3). Because of the large number of repeat sequences in the mammalian genome, approximately 80% of the clones should give a signal after hybridization to this probe if the library was generated in a completely unbiased manner.

Protocol 4.1. Assay for Repeat Sequences (19, 26)

1. Select about 50 clones randomly and purify DNA inserts as described earlier (Protocol 1.1 in this chapter).
2. Subject DNA to electrophoresis on agarose gel and blot onto nitrocellulose filter (1).
3. Radioisotopically label total genomic DNA or cloned repeat core sequences, for example, clones containing *Alu* sequences such

as Blur-8 (BRL) (26), with [^{32}P]dATP by nick translation (27). Hybridize labeled probe to filters at temperatures that are adjusted to a stringency that favors the hybridization of repeat sequences (42–48°C) (26).

4. Divide the number of clones that give a positive signal by the total number of clones transferred to the filter to determine the percentage of clones containing repeat sequences.

Protocol 4.2. Assay for Unique Sequences

1. Digest total genomic DNA with several restriction enzymes (e.g., BamHI and PstI) and transfer to membrane filters for hybridization (1).

2. Hybridize radioisotopically labeled recombinant DNA inserts from separate clones to Southern blots under conditions that favor unique sequences (65–68°C) (1).

3. Wash filters sequentially in 2× SSC, 0.5% SDS, then in 2× SSC, 0.1% SDS, to a final stringency of 0.1× SSC, 0.1% SDS at 65°C. Hybridization temperatures are near 55°C when formamide is used in the hybridization mix (1).

From the number of clones analyzed, determine

a. percentage of those containing repeat sequences (hybridized to total genomic DNA)

b. percentage of inserts containing unique sequences (hybridized to a single or small subset of restriction enzyme fragment bands on the Southern blots)

Some DNA inserts may give no hybridization signal, because of inadvertent cloning of DNA sequences not in the dissected chromosomal region, cloning from chromosomal DNA degraded by excessive acid fixation and aging of chromosomes, or cloning of contaminating DNA (e.g., bacterial DNA on the slide).

To illustrate these calculations, the following hypothetical example is used:

A chromosomal band-specific recombinant library was constructed. The library was estimated to have 5000 clones, 50 independent clones of which were selected randomly to be analyzed.

The insert size ranged between 100 and 500 bp with a mean average size of 217 bp.

Of the selected clones, 22 (44%) gave no signal in Southern hybridization to genomic DNA.

Of the selected clones, 9 (18%) contained repeat sequences.

Thus, the number of clones with unique sequences (i.e., those that may encode structural genes) was 19 (38%) and the total number of cloned sequences (repeat plus unique sequences) was 28 (56%).

Calculation of the Percentage of Total Microdissected DNA Cloned

An ideal library generated from a chromosome fragment should consist of a series of recombinant DNA clones that contain overlapping sequences that cover the entire chromosomal region. The size of the library of completely random fragments of genomic DNA necessary to ensure representation of a particular sequence of interest is dictated by the size and number of the cloned inserts and the size of the chromosomal fragment originally dissected.

Theoretically, the number of clones required to achieve a library with a 99% probability of containing a particular sequence is determined by the following formula (28):

$$N = \ln(1 - P)/\ln[1 - (I/G)],$$

where P is the desired probability, I is the insert size, G is the genome size of the fragment, and N is the necessary number of clones. Therefore, if one generates a library with an average insert size of 2.4 kb (the average length of *Eco*RI digestion fragments), the number of clones necessary to ensure a 99% probability of cloning the entire 10 Mbp in a dissected fragment is determined as:

$$N = \ln(1 - 0.99)/\ln[1 - (2.4 \times 10^3/10^7)] = 19,000.$$

This calculation represents the number of hypothetical clones required. In general, to have a 99% chance of isolating a desired sequence, the number of clones screened should be great enough that the total number of base pairs present in the clones screened ($I \times N$) represent a 4.6-fold excess over the total number of base pairs in the genome (G) (29).

The actual number of clones obtained is determined by factors such as the efficiency of the cloning method, the cloning capacity of the vector, the relative number and size of clonable and nonclonable DNA sequences in the target DNA, and the amount of DNA degradation in the process of chromosome preparation and library construction. These factors may explain why only a small fraction of microdissected DNA sequences is represented in libraries constructed to date (5–12, 30). The following guidelines help determine the cloning efficiency from mammalian chromosomal DNA (31, 32).

1. Estimate the amount of DNA recovered from microdissection for microcloning. Assume that a mammalian chromosome fragment representing an average size GTG band has a length of 10,000 kb (33).

2. Determine the number of clones obtained in your recombinant library.

3. Pick 50 clones randomly for analysis. Compare the average size of clones with the average size that is determined empirically from the analysis of the digestion of genomic DNA with a specific restriction enzyme. Determine the frequency of redundant clones. Determine the number of independent cloned repeat and unique sequences.

4. Determine the percentage of the total DNA cloned using the formula:

% fragment DNA cloned = (number of cloned sequences × mean average insert size)/(fragment DNA content) × 100.

In the example we discussed earlier, 28 clones were positive in Southern blot hybridization. These clones represent 56% of those analyzed. Therefore, the estimated number of cloned sequences in the library is 5000 × 0.56 = 2800. Assuming that the clones in the library are distributed randomly along the fragment with an average spacing of 10 kbp, the percentage of fragments cloned would be

(2800 × 217 bp/10,000 kbp) × 100 ~ 6.0%.

This number represents the percentage of chromosomal fragments cloned.

Determination of Potential Structural Gene Sequences

Evolutionarily conserved DNA sequences are indicative of potential structural gene sequences. The relatedness of a defined recombinant sequence among various species is identified readily by its hybridization to genomic DNA from these species using relatively stringent hybridization conditions. Genomic DNA from a number of species is now commercially available on biomembrane blots known as the Biomap Evo blots (Biosmap products, BIOS), sometimes also referred to as "Zooblots." These blots contain genomic DNA from the human, cat, dog, sheep, pig, cow, rabbit, rat, and mouse. Potential genes are identified as those that hybridize to DNA sequences from at least three species (30). Once that ability is determined, such sequences can be used to isolate the full-length expressed gene from cDNA libraries of the appropriate tissue. Structural genes may be identified by hybridizing cloned genomic DNA to cDNA libraries under conditions that suppress hybridization of repeat sequences (34).

Localization of Recombinant Clones Using
in Situ Hybridization

After performing the detailed analysis of the recombinant clones that we outlined earlier, *in situ* hybridization must be performed to confirm whether the clones map to the microdissected band. *In situ* hybridization permits the mapping of specific DNA sequences to individual chromosomes with excellent regional localization. In addition to identifying the chromosomes on which a gene lies, fluorescence *in situ* hybridization also identifies the region of the chromosome to a resolution of about $1-2 \times 10^7$ bp (0.5% of the genome) (35, 36). Specific chromosomal subregions can be highlighted specifically and localized in metaphase or interphase nuclei using radioactive or fluorescent labeled probes (35–37).

For the purpose of complete characterization of a chromosome band-specific library, we will discuss only fluorescent *in situ* hybridization techniques (FISH), since these techniques allow visualization of targeted genes more readily and with higher spatial resolution than can be obtained using radioisotopically labeled *in situ* hybridization probes. Further, the signal may be amplified by at least two orders of magnitude using antibodies specific to the reporter molecule. Thus, signal resolution is improved to the extent of localizing a sequence as small as 1000 kbp (38–42). Probes tagged with different fluorophores can be used, thus permitting simultaneous detection of more than one sequence in the same preparation (43). Unique sequences contained in large insert probes, such as cosmid and YAC clones, can be used directly in FISH after suppression hybridization of repetitive sequences with unlabeled genomic DNA or repeat sequences (44).

For discussions of theoretical and historical background, sensitivity of the technique, choice of probe labeling, and chromosome banding techniques, we recommend the excellent manuals by Malcolm *et al.* (35) and Polak *et al.* (36).

Reagents and Materials

Carnoy's fluid (10 : 60 : 30 glacial acetic acid:ethanol:chloroform)
Chromosome preparation and staining reagents (see Chapter 2)
Cot1 DNA (BRL)
Coverslip sealant (Oncor Corp.)
Earle's solution (GIBCO, BRL)
EDTA, 2 mM

Ethanol, cold (70%, 80%, 95%, 100%)
Fluorescence labeling kit (Oncor Corp.; Boehringer-Mannheim Biochemicals; Imagenetics, Inc.)
Fluorescence microscope equipped with fluorescein isothiocyanate (FITC) and propidium iodide/rhodamine fluorescence detection optical filters (450–490 nm/ 515 nm)
Glass coverslips, sterile
Humidified chamber
Hybridization buffer—50% formamide, 10% dextran sulfate, 0.025 M NaCl, 0.04 M sodium phosphate, 0.03 M sodium citrate (pH 6.0), 0.5 × Denharot's solution, 200 μg/ml denatured salmon sperm DNA
Incubator, 37°C
Microcentrifuge tubes
Micropipettes and micropipette tips
Microscope slides, standard, sterile
Nick translation kit (random prime kit may be used) (Boehringer-Mannheim)
Nylon membrane
Phosphate-buffered saline (PBS) (GIBCO, BRL)
Proteinase K solution—1 μg/ml in 20 mM Tris–HCl (pH 7), 2mM CaCl$_2$
RNase stock solution—10 mg/ml in 2× SSC; boil for 3 minutes and cool rapidly to inactivate any contaminating DNase; working solution is 100 μg/ml in 2× SSC.
Sephadex G-50 or BioGel P-2 column
Slide warmer (Fisher Scientific)
SSC buffer—0.15 M NaCl, 0.015 M sodium citrate
Water, sterile, deionized
Water bath, 15°C, 70°C

DNA Probe Preparation

The fluorescent labeling system is based on the modification of a DNA probe with the hapten biotin. Biotin is detected by binding to avidin or streptavidin that is coupled to enzymes such as horseradish peroxidase or to fluorochromes. Significant amplification of the signal can be achieved using a second layer of a biotinylated anti-avidin antibody and a third layer of fluorochrome-coupled avidin (35, 36).

The DNA to be used for *in situ* hybridization should be of the highest quality. Purification through a CsCl gradient is preferred. Cleaving the probing sequence from the vector is not necessary. The extra DNA contained in the vector may provide additional networks during hybridization. The probe should be devoid of repetitive sequences. If any repeat sequences are present, they must be removed first by subcloning into a suitable vector or by using suppression hybridization in the detection reaction.

Nick translation is the commonly accepted method for labeling probe DNA (27). All the components for the reaction are readily available in kits from biochemical companies that specialize in molecular biology products. Care should be taken to optimize the reaction conditions to give the maximum incorporation of biotinylated dUTP. Figure 4.4 illustrates the essential steps of fluorescence in *in situ* hybridization. A detailed procedure follows.

Protocol 5.1. DNA Probe Labeling for Fluorescence in Situ Hybridization

1. To a microcentrifuge tube, add
    a. 1 μg candidate probe DNA
    b. 10 μl nucleotide mix (with biotin-labeled dUTP)
    c. 5 μl 10× nick translation enzyme mixture (Boehringer-Mannheim Biochemicals)
2. Adjust to 50 μl with DNase-free deionized water.
3. Mix well and spin for 2 seconds in a microcentrifuge.
4. Place the tube in a 15°C water bath for 75 minutes.
5. Prepare a Sephadex G-50 or BioGel P-2 column according to the instructions of the manufacturer.

Molecular Cloning 113

```
                    Prepare metaphase chromosomes and
                    immobilize karyotype on microscope cover glass
                                    ↓
                    Perform RNase treatment on chromosomal DNA
                    10 µg/ml/1 hr/37°C
                                    ↓
                    Wash 3 times with 2X SSC
                                    ↓
                    Dehydrate DNA in cold ethanol (70%, 80%, 95%), 2 minutes each
                                    ↓
                    Denature target DNA at 70°C in 2X SSC/70% formamide
                                    ↓
Label probe with biotin
using nick translation or    →      Incubate slides with
mix probe with hybridization        probe at 37°C for 16 hr
mix and denature at 70°C
                                    ↓
                    Wash slides sequentially in 2X SSC (43°C), 2X SSC (37°C),
                    1X SSC (42°C)
                                    ↓
                    Incubate with fluorescein-conjugated
                    avidin, 20 minutes at 37°C, for
                    detection of hybridization signal
                                    ↓
                    Amplify hybridization signal by
                    addition of anti-avidin antibody and
                    FITC–avidin, 20 minutes at 37°C
                                    ↓
                    Counterstain slides with DAPI or
                    propidium iodine (slides may be stained with Giemsa stain to
                    generate GTG banding prior to hybridization)
                                    ↓
                    Mount slides in glycerol based
                    medium with antifading agent (e.g., DABCO) for viewing and
                    photography using 400 ASA film and the appropriate optical
                    filter set
```

Figure 4.4 Flow chart of fluorescence *in situ* hybridization with biotinylated probe.

6. Apply the DNA probe sample to the column and centrifuge at 175 g for 5 minutes. Collect the effluent in the microcentrifuge. (Unincorporated nucleotides will remain trapped in the column.)

7. Adjust the final volume of biotinylated DNA to 100 μl with 2 mM EDTA (DNase-free).

8. Prepare serial dilutions (1:10, 1:100, and 1:1000) of the biotinylated probe and of the control DNA (usually provided by the manufacturer). On a small strip of nylon membrane, mark two rows with spots of each dilution of both DNAs. Bake the membrane for 30 minutes at 80°C. Determine the level of biotin incorporated into the probe using the detection assay described next. Calculate the concentration of labeled probe using the dilution that gives a fluorescent signal that closely matches a control DNA sample.

Protocol 5.2. Fluorescent in Situ Hybridization to Metaphase Chromosome Spreads

This procedure involves two steps: (1) denaturing the two strands of the chromosomal DNA contained in the metaphase spread on a coverslip, followed by hybridization to a denatured fluorescent-labeled oligonucleotide probe, and (2) localization of the hybridization signal to a specific chromosomal band region.

Preparation of Slides for in Situ Hybridization

1. Prepare metaphase chromosome spreads or interphase nuclei on sterile clean microscope coverslips or slides according to procedures described in Chapter 2 of this manual. Baking slides overnight at 56°C is not recommended. Heat denaturation reduces the fluorescence signal. Keep slides at -20°C in an air-tight box until ready to use.

2. Fix chromosomes on the slides using Carnoy's fluid (10:60:30 glacial acetic acid:absolute ethanol:chloroform) for 20 minutes.

Carnoy's is an excellent fixative and increases the retention of chromosomes on the slide throughout the following steps.

3. Dehydrate slides by dipping in 70% ethanol (twice) and in 95% ethanol once. This step is necessary to minimize the background over the chromosomes and to ensure adherence of chromosomes to the slides throughout the procedure.

RNase Treatment

1. Incubate the slides for 1 hour with 200 μl RNase at 37°C. This step is necessary to remove any endogenous hybridizable RNA.

2. Wash the slides thoroughly three or four times in 2× SSC (2 minutes each) at room temperature.

Dehydration

1. Dehydrate slides in a series of cold 70%, 80%, and 95% ethanol (2 minutes each).

2. Allow slides to air dry for 10–15 minutes.

Denaturation

1. Incubate slides for 2 minutes at 70°C in 2× SSC solution containing 70% formamide.

2. Immediately transfer slides to ice-cold 70% ethanol to reduce strand reannealing before adding the DNA probe. Repeat rinsing in cold 80%, 90%, and 100% ethanol solutions.

3. Allow slides to air dry for a few minutes.

Hybridization

1. Prewarm slides to approximately 37°C in an incubator or on a slide warmer.

2. Digest chromosomal proteins by incubating with proteinase K solution for 5–8 minutes at 37°C. Follow with dehydration procedure (Steps 1 and 2 under Dehydration).

3. Prepare hybridization mixture by mixing 1.5 μl biotin-labeled DNA probe (10 ng/μl) with 28.5 μl hybridization buffer for each slide in an Eppendorf tube. The amount of DNA probe may vary from 10 to 50 ng per slide.

4. Denature probe by heating in a 70°C water bath for 5 minutes. Quickly chill on ice.

5. Add 30 μl mixture directly to each slide and cover with a coverslip or plastic slide. Seal cover with coverslip sealant (normally provided with kit).

6. Incubate for 16 hours at 37°C. Depending on the concentration and complexity of the probe, hybridization time may vary from 2 hours to overnight at 37–45°C. Shorter times are used for detection of abundant sequences, whereas longer times are required for detection of unique DNA sequences. Prehybridization of the probe mixture with excess unlabeled genomic or Cot1 DNA for up to 1 hour (suppression hybridization) is necessary to reduce the diffuse hybridization of repetitive sequences in the probe to multiple chromosomal sites (16, 19, 20, 39).

Postwashing

1. Carefully remove the coverslip sealant with forceps. **Do not** remove the coverslips.

2. Place slides in 2 × SSC at 43°C for 20 minutes with gentle agitation. The coverslip will come off under these conditions.

3. Wash slides in 2 × SSC solution for 20 minutes at 37°C.

4. Repeat washing once at 42°C in excess 1 × SSC and once at room temperature in 0.1 × SSC (4 minutes each). Note that wash-

ing temperatures may vary depending on the complexity of the probe.

Detection

1. Do not allow the slides to dry after the washing steps.
2. Incubate each slide in 60 µl nonspecific serum for 5 minutes at room temperature.
3. Apply 60 µl fluorescein-labeled avidin (5µg/ml; Vector Laboratories) to each slide to produce a fluorescent signal at the sites of probe hybridization. Incubate 20 minutes in humidified chamber or on a slide warmer at 37°C.
4. Carefully drain reagent off the slide.
5. Wash slides three times in 1× PBS at room temperature, for 2 minutes each wash.

Signal Amplification

The hybridization signal may be amplified once with additional layers of goat anti-avidin antibody (Vector Laboratories) and fluorescein-labeled avidin as described elsewhere (42, 45).

1. Carefully remove coverslip and allow liquid to drain off slide.
2. Apply 60 µl anti-avidin antibody (5µg/ml) to each slide and replace plastic coverslip. Dilutions of antibodies are determined empirically. A dilution in the range of 1:40 to 1:70 normally works well.
3. Incubate for 20 minutes in a humidified chamber at 37°C.
4. Wash slides three times in excess 1× PBS at room temperature, for 2 minutes each wash.
5. Repeat detection steps 2–4 above. For detection of single copy

genes, the amplification step may be repeated an additional time before the fluorescent signal becomes clearly visible.

Chromosome Staining and Identification

To identify specific chromosomes and chromosomal bands, one of the following four methods may be used.

Method 1. GTG Banding Followed by FISH and Propidium Iodide Fluorescence Staining (46)

1. Perform GTG banding as described by Seabright (47). Also, see Chapter 2 for details.
2. Find, identify, and mark chromosomal spreads and the region of interest.
3. Photograph spreads.
4. Destain chromosomes on slides with 75% ethanol; air dry briefly.
5. Proceed with FISH procedure.
6. Compare FISH results with GTG-banded chromosomes

Method 2. FISH Hybridization Followed by R Banding Using Hoechst Dye (45)

1. Perform FISH first.
2. Stain chromosomes with Hoechst 33258 dye (0.1 µg/ml) for 15 minutes.
3. UV irradiate slides for 20 minutes at 365 nm.
4. Immerse slides in Earle's solution (pH 6.5) at 87°C for 2 minutes.
5. Counterstain chromosomes with propidium iodide (1 µg/ml in antifade solution, DABCO).

Method 3. Simultaneous FISH and R Banding Using Propidium Iodide and DAPI (48)

1. After the FISH procedure, the slides are washed and mounted in medium that contains DAPI (4,6-diamidino-2-phenylindole; 0.8 µg/ml) and propidium iodide (0.4 µg/ml).

2. To visualize an *in situ* signal stain, add 25 µl glycerol-based medium (e.g., Immuno mount, Shandon, or Aqueous mounting medium; Biomedia, Inc.) containing 2.5% 1,4-diazobicyclo-2,2,2-octane (DABCO) to each slide, cover with a coverslip, and view under a fluorescence microscope. DABCO is used as an antifading agent to minimize photobleaching during microscopic examination of the slides.

3. The following optical filter sets are used for visualization of hybridization signals and R bands:

    a. Use excitation filter combination of 450–490 nm/515 nm barrier to observe apple green spots (FITC probe) on red chromosomes (propidium iodide). With this filter combination, the red emission ranges are blocked and only the fluorescence FITC is seen.

    b. Use a selective filter of 515–560 nm/580 nm barrier to visualize the fluorescent R bands.

4. Take photomicrographs using Kodak Ektachrome 400 ASA film. For better photographs, double exposure of the FITC signal or fluorescence R banding is recommended.

5. See Lichter *et al.* (49) and Harnden and Klinger (50) for excellent guides to the identification of human chromosomes and banding patterns.

Figure 4.5 illustrates FITC-labeled FISH probe generated using Method 3 and hybridized to a normal lymphocyte metaphase chromosomes, then stained with propidium iodide.

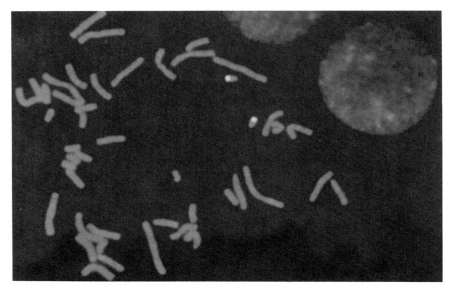

Figure 4.5 Fluorescence *in situ* hybridization of probes generated from microdissected chromosomal DNA with unidentifiable translocation in a human metaphase spread. [Adapted from Meltzer *et al.* (16).]

Method 4. Chromosome Painting

Polymerase chain reaction has been combined with chromosome flow sorting to generate pools of DNA sequences that map to specific chromosomes. Such probes were used in fluorescence *in situ* hybridization to identify regions of normal chromosomes involved in translocations (51–53). In addition, these probes can be used to highlight specific sequences that span the length of the target chromosome in a karyotype (chromosome painting). Since the chromosome of interest in microdissection and cloning experiments is known, a quick alternative to GTG or fluorescence banding in identifying that chromosome in *in situ* hybridization is the use of chromosome painting probes. Painting probes are now readily available from Oncor (Bio-probes) and GIBCO BRL (SpectrumGreen). The Oncor Painting Palette includes probes for all human chromosomes. Painting probes and test probes with different fluorescence labels are premixed with blocking DNA and appropriate hybridization solution, thus allowing instant identification of chromosomes without banding procedures.

Troubleshooting and Solutions

1. *Chromosomes do not band well enough to identify chromosomes.*

    emperically determine optimal conditions for specific cells and chromosomes used; try different banding or staining procedure (e.g., R banding vs. GTG banding)
    GTG banding normally is done before FISH to avoid distortion of chromosomal bands by FISH procedure

2. *Hybridization signal is weak or not visible.*

    try labeling a different clone from the library to use as a probe; increase the specific activity or use a larger probe; the effective concentration of the probe per slide (either actual probe concentration or reaction volume) will contribute significantly to signal detection
    add a second layer of anti-avidin antibody and fluoresceinated avidin to increase signal detection
    add dextran sulfate (10%) to hybridizations to increase probe–chromosome network formation
    overstaining with the counterstaining dye may quench or diminish the hybridization signal

3. *Signal is visible but background is high.*

    this problem is a function of optical filters and barrier filter combination set used; the narrower the band width of the excitation filter, the better, since it minimizes the emission from the dye used for counterstataining
    use of a confocal microscope, if available, helps block any background from off-focus planes since one can focus on the particular plane in which the fluorescent signal is located

4. *Background over chromosomes is still high.*

    remove repeat sequences by subcloning the unique sequence portion of the probe into a suitable vector

    remove nonspecific hybridization by suppression hybridization with Cot1 DNA or genomic DNA

    adjust posthybridization washing conditions to be more stringent (higher temperature, lower salt concentration)

    use high quality optics with high resolution (high numerical aperture) or a confocal microscope to improve signal significantly

5. *Signal is visible but good photomicrographs are unobtainable.*

    problem a function of filters and barrier filters used to visualize the fluorescent signal; high speed film (400 ASA) and double exposure help increase signal to noise ratio

    image enhancement systems (e.g., Noran Odyssey Image 1) permit manipulations such as subtracting backgrounds, increasing contrast, frame averaging, image stacking, and enhancement

Further Characterization

Once the potential clones in the minilibrary are characterized as described they are ready for further analysis:

1. Subclone into an appropriate plasmid vector (e.g., Bluescribe or pUC 13).

2. Generate an extensive restriction enzyme map of cloned DNA inserts.

3. Perform direct sequencing.

4. Use labeled DNA clone to isolate full-length gene from a cDNA or genomic library.

References

1. Sambrook, J., Fritsch, E. F., and Maniatis, T. (1989). "Molecular Cloning: A Laboratory Manual," Volumes 1–3. Cold Spring Harbor Laboratory Press, Cold Spring Harbor, New York.

2. Davis, L. G., Dibner, M. D., and Battey, J. F. (1986). "Basic Methods in Molecular Biology." Elsevier Science Publishing, New York.

3. Ausubel, F. M., Brent, R., Kingston, R. K., Moore, D. D., Seidman, J. G., Smith J. A., and Struhl, K. (1987). "Current Protocols in Molecular Biology." Greene Publishing Associates and Wiley–Interscience, New York.

4. Scalenghe, F., Truco, E., Edström, J. E., Pirrota, V., and Melli, M. (1981). Microdissection and cloning of DNA from specific region of *Drosophila melanogaster* polytene chromosomes. *Chromosoma (Berlin)* **82**, 205–216.

5. Röhme, D., Fox, H., Herrmann, H., Frischauf, A.-M., Edström, J. E., *et al.* (1984). Molecular clones of the mouse t complex derived from microdissected metaphase chromosomes. *Cell* **36**, 783–788.

6. Fisher, E. M., Cavanna, J. S., and Brown, S. D. M. (1985). Microdissection and microcloning of the mouse X chromosome. *Proc. Natl. Acad. Sci. USA* **82**, 5846–5849.

7. Bates, G. B., Wainwright, B. J., Williamson, R., and Brown, S. M. (1986). Microdissection and microcloning from the short arm of human chromosomes 2. *Mol. Cell. Biol.* **11**, 3826–3830.

8. Kaiser, R., Webber, J., Greezechik, K.-H., Edström, J. E., *et al.* (1987). Microdissection and microcloning of the long arm of human chromosome 7. *Mol. Biol. Rep.* **12**, 3–6.

9. Lüdecke, H-J., Senger, G., Claussen, U., and Horsthemke, B. (1989). Cloning defined regions of the human genome by micro-

dissection of banded chromosomes and enzymatic amplification. *Nature (London)* **338**, 348–350.

10. Senger, G., Lüdecke, H.-J., Horsthemke, B., and Claussen, U. (1990). Microdissection of banded human chromosomes. *Hum. Genet.* **84**, 507–511.

11. Lüdecke, H. J., Senger, G., Claussen, U., and Horsthemke, B. (1990). Construction and characterization of band-specific DNA libraries. *Hum. Genet.* **84**, 512–516.

12. Johnson, D. (1990). Molecular cloning of DNA from specific chromosomal regions by microdissection and sequence-independent amplification of DNA. *Genomics* **6**, 243–251.

13. Horsthemke, B., Claussen, U., Hesse, S., and Lüdecke, H-J. (1992). PCR mediated cloning of *Hpa*II tiny fragments from microdissected human chromosomes. *PCR Meth. Appl.* **1**, 299–233.

14. Wesley, C. C., Ben, M., Kreitman, M., Hagag, N., and Eanes, W. (1990). Cloning regions of the *Drosophila* genome by microdissection of polytene chromosome DNA and PCR with nonspecific primer. *Nucleic Acids Res.* **18**, 599–603.

15. Wesley, U., Hagag, N., and Viola, M. (1991). Construction of chromosome band specific minilibrary using microdissection and PCR amplification. Unpublished results.

16. Meltzer, P. S., Guan, X.-Y., Burgess, A., and Trent, J. M. (1992). Rapid generation of region specific probes by chromosome microdissection and their application. *Nature Genetics* **1**, 24–28.

17. Feinberg, A. P., and Vogelstein, B. (1984). A technique for radiolabeling DNA restriction endonuclease fragments to high specific activity. *Anal. Biochem.* **137**, 266–267.

18. Nelson, D. L., Ledbetter, S. A., Corbo, L., Victoria, M. F., Ramirez-Solis, R., Webster, T. D., Ledbetter, D. H., and Caskey, T. (1989). *Alu* polymerase chain reaction: A method for rapid

isolation of human-specific sequences from complex DNA sources. *Proc. Natl. Acad. Sci. USA* **86**, 6686–6690.

19. Lengauer, C., Reithman, H., and Cremer, T. (1990). Painting of human chromosomes with probes generated from hybrid cell lines by PCR with *Alu* and L1 primers. *Hum. Genet.* **86**, 1–6.

20. Ledbetter, S. A., Nelson, D. L., Warren, S. T., and Ledbetter, D. H. (1990). Rapid isolation of DNA probes within specific chromosomes regions by interspersed repetitive sequence (IRS) PCR. *Genomics* **6**, 475–481.

21. Bicknell, D. C., Markie, D., Spurr, N. K., and Bodmer, W. F. (1991). The human chromosome in human × rodent somatic cell hybrids analyzed by a screening technique using *Alu* PCR. *Genomics* **10**, 186–192.

22. Breen, M., Arveiler, B., Murray, I., Godson, J. R., and Porteous, D. J. (1992). YAC mapping by FISH using alu–PCR-generated probes. *Genomics* **13**, 726–730.

23. Nelson, D., Ballabio, A., Victoria, M. F., Pieretti, M., Bies, R., Gibbs, R. A., Maley, J., Chinault, C., and Webster, T. D. (1991). Alu-primed polymerase chain reaction for regional assignment of 110 yeast artificial chromosome clones from the human X chromosome: Identification of clones associated with a disease locus. *Proc. Natl. Acad. Sci. USA* **88**, 6157–6161.

24. Cole, C. G., Goodfellow, P. N., Bobrow, M., and Bentley, D. R. (1991). Generation of novel sequence tagged sites (STSs) from discrete chromosomal regions using alu PCR. *Genomics* **10**, 816–826.

25. Griffin, L. D., McGregor, G. R., Muzny, D. M., Hater, J., Cook, R. G., and McCabe, E. R. B. (1988). Synthesis of hexokinase 1 cDNA probes by mixed oligonucleotide-primed amplification of cDNA (MOPAC) using primer mixtures of high complexity. *Am. J. Hum. Genet.* **43**, A185.

26. Nisson, P. E., Watkins, P. C., Menninger, J. C., and Ward D. (1991). Improved suppression hybridization with human DNA (Cot-1 DNA) enriched for repetitive DNA sequences. *Focus* **13**, 42–45.

27. Rigby, P. W. J., Diechmann, M., Rhodes, C., and Berg, P. (1977). Labelling deoxyribonucleic acid to high specific activity *in vitro* by nick translation with DNA polymerase I. *J. Mol. Biol.* **113**, 237–251.

28. Clark, L., and Carbon, J. (1976). A colony bank containing synthetic ColE1 hybrids representative of the entire *E. coli* genome. *Cell* **9**, 91–99.

29. Seed, B., Parker, R. C., and Davidson, N. (1982). Representation of DNA sequences in recombinant DNA libraries prepared by restriction enzyme partial digestion. *Gene* **19**, 201–209.

30. Buiting, K., Neumann, M., Lüdecke, H.-J., Senger, G., Claussen, U., Antich, J., Parsage, E., and Horsthemke, B. (1990). Microdissection of Prader–Willi syndrome chromosome region and identification of potential gene sequences. *Genomics* **6**, 521–527.

31. Amar, L. C., Arnaud, D., Cambrou, J., Gueneet, J.-H., and Avner, R. R. (1985). Mapping of the mouse X chromosome using random genomic probes and an interspecific mouse cross. *EMBO J.* **4**, 3695–3700.

32. Brown, S. D. M., and Greenfield, A. J. (1987). A model to describe the size distribution of mammalian genomic fragments recovered by microcloning. *Gene* **55**, 327–332.

33. Röhme, D. (1986). Microdissection of metaphase chromosomes and gene transfer. *In* "Microdissection and Organelle Transplantation: Methods and Applications" (J. E. Celis, A. Graessmann, and A. Loyter, eds.). Academic Press, New York.

34. Lovett, M., Kere, J., and Hinton, L. M. (1991). Direct selection: A method for the isolation of cDNAs encoded by large genomic regions. *Proc. Natl. Acad. Sci. USA* **88,** 9628–9632.

35. Malcolm, S., Cowell, J. K., and Young, B. D. (1986). Specialist techniques in research and diagnostic clinical cytogenetics. In "Human Cytogenetics: A Practical Approach" (D. E. Rooney and B. H. Czepulkowski, eds.). IRL Press, Oxford.

36. Polak, J. M., and McGee, J. O. (1990). In "*In Situ* Hybridization: Principles and Practice." Oxford University Press, New York.

37. Trask, B. J., van Der Engh, G., Pinkel, D., Mullikin, J., Waldman, F, Van Pekken, H., and Gray, J. (1988). Fluorescence *in situ* hybridization to interphase cell nuclei in suspension allows flow cytometric analysis of chromosome content and microscopic analysis of nuclear organization. *Hum. Genet.* **78,** 251–259.

38. Trask, B. J. (1991). Fluorescence *in situ* hybridization: Applications in cytogenetics and gene mapping. *Trends Genet.* **7,** 149–155.

39. Lichter, P., Ledbetter, S., Ledbetter, D. H., and Ward, D. C. (1990). Fluorescence *in situ* hybridization with *Alu* and L1 polymerase chain reaction probes for rapid characterization of human chromosomes in hybrid cell lines. *Proc. Natl. Acad. Sci. USA* **87,** 6634–6638.

40. Baldini, A., and Ward, D. C. (1991). *In situ* hybridization banding of human chromosomes with *Alu*–PCR products: A simultaneous karyotype for gene mapping studies. *Genomics* **9,** 770–774.

41. Lawrence, J. B., Villanave, C. A., and Singer, R. H. (1988). Sensitive, high resolution chromatin and chromosome mapping *in situ:* Presence and orientation of two closely integrated copies of EBV in a lymphoma line. *Cell* **52,** 51–62.

42. Pinkel, D., Landegent, J., Collins, C., Fuscoe, J., Segraues, R., Lucas, J., and Grey, J. (1988). Fluorescence *in situ* hybridization

with human chromosome specific libraries: Detection of trisomy 21 and translocations of chromosome 4. *Proc. Natl. Acad. Sci. USA* **85,** 9138–9142.

43. Nederlof, P. M., van der Flier, S., Wiegant, J., Raap, A. K., Tanke, H. J., Ploem, J. S., and van der Ploeg, L. (1990). Multiple fluorescence *in situ* hybridization. *Cytometry* **11,** 126–131.

44. Landegent, J. E., Jansen, N., Dewal, I. N., Pirks, R. W., Baas, F., and van der Ploeg, M. (1987). Use of whole cosmid cloned genomic sequences for chromosomal localization by nonradioactive *in situ* hybridization. *Hum. Genet.* **77,** 366–370.

45. Cherif, D., Julier, C., Delattre, O., Derre, J., Lathrop, G. M., and Berger, R. (1990). Simultaneous localization of cosmids and chromosome R banding by fluorescence microscopy: Application to regional mapping of human chromosome 11. *Proc. Natl. Acad. Sci. USA* **87,** 6639–6643.

46. Smit, V. T. H. B. M., Wessels, J. W., Mollevanger, P., Dauwerse, J. G., van Vliet, M., Beverstock, G. C., Breuning, M. H., Devilee, P., Raap, A. C., and Cornelisse, C. J. (1991). Improved interpretation of complex chromosomal rearrangements by combined GTG banding and *in situ* hybridization using chromosome specific libraries and cosmid probes. *Genes Chrom. Cancer* **3,** 239–248.

47. Seabright, M. (1971). A rapid banding technique for human chromosomes. *Lancet* **ii,** 971–972.

48. Fan Y.-S., Davis, L. M., and Shows, T. B. (1990). Mapping small DNA sequences by fluorescence *in situ* hybridization directly on banded metaphase chromosomes. *Proc. Natl. Acad. Sci. USA* **87,** 6223–6227.

49. Lichter, P., Boyle, A. L., Cremer, T., and Ward, D. C. (1991). Analysis of genes and chromosomes by nonisotopic *in situ* hybridization. *Genetic Anal. Techniques Appl.* **8,** 24–35.

50. Harnden, D. G., and Klinger, H. P. (eds.) (1985). "An International System for Cytogenetic Nomenclature." Karger, Basel.

51. Telenius, H., Pelmear, A. H., Tunnacliffe, A., Carter, P. N., Behmel, A., Ferguson-Smith, M., Nordenskjold, M., Pfragner, R., and Ponder, B. A. J. (1992). Cytogenetic analysis by chromosome painting using DOP–PCR amplified flow sorted chromosomes. *Genes Chrom. Cancer* **4**, 257–263.

52. Lengauer, C., Eckelt, A., Weith, A., Endlich, N., Ponelies, N., Lichter, P, Greulich, K. O., and Cremer, T. (1991). Painting of defined chromosomal regions by *in situ* suppression hybridization of libraries from laser-microdissected chromosomes. *Cytogenet. Cell Genet.* **56**, 27–30.

53. Chang, K-S., Vyas, R. C., Deaven, L. L., Trujillo, J. M., Stass, S. A., and Hittelman, W. N. (1992). PCR amplification of chromosome-specific DNA isolated from flow cytometry-sorted chromosomes. *Genomics* **12**, 307–312.

Chapter 5

Applications of Chromosome Microdissection

In this chapter we review the potential uses of microdissected chromosome fragments once they have been collected. These applications are discussed in three general experimental categories:

1. direct analysis of the PCR product of microdissected chromosome fragments
2. recombinant DNA libraries generated from microdissected chromosome fragments
3. gene transfer using chromosome fragments

Direct Analysis of the PCR Product of Microdissected Chromosome Fragments

Gene Mapping

Methods used to localize a known gene to a specific chromosome band (e.g., hybridization of the gene to rodent–human hybrid cells containing portions of human chromosomes, *in situ* hybridization to metaphase spreads) may yield ambiguous results or may be able only to localize a gene to a broad chromosome region. If the sequence of the gene of interest is known, oligodeoxynucleotide primers may be synthesized and used in a PCR diagnostic for that gene. The ability of DNA from a microdissected band to template PCR using gene-specific primers localizes that gene to that specific chromosome region (1).

Mapping Sites of Chromosome Rearrangements and Deletions

If the spatial relationship of multiple genes in a specific chromosome region is known, amplification of target sequences using PCR can be used to detect structural alterations (deletions or rearrangements). A limited number of chromosome fragments from one region can be dissected and target sequences then amplified using site-specific oligonucleotide primers. The inability to generate a PCR product using DNA from microdissected chromosome fragments (in the presence of appropriate positive controls) indicates that the target sequences are not present (deleted or rearranged). In Figure 5.1, a translocation site was microdissected. The analysis of the PCR products indicates that the translocation site is between target sequences B and C in the microdissected fragment(s). Using this method, Cotter *et al.* (1) have mapped the breakpoints on chromosome 11 that are present in a number of translocations associated with human leukemia.

When a translocation is identified by karyotype analysis but the chromosomal origin of the derivative chromosomes is ambiguous, analysis of the microdissected translocation site may be helpful (2, 3). One strategy involves microdissection of the region of translocation, followed by *in vitro* amplification (PCR) using a nonspecific oligonucleotide primer containing a stretch of degenerate bases (2). The PCR product is fluorescence labeled and hybridized to a normal karyotype to identify the chromosomes that participate in the translocation. This method (called micro-FISH) is rapid and may expedite the identification of complex karyotypes.

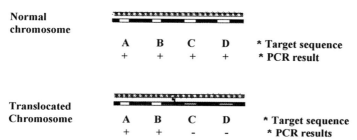

Figure 5.1 Use of microdissection and PCR to map chromosome rearrangements.

Determination of Coupling Phase

Knowledge of the linkage of specific alleles of two different genes (coupling phase) on the same chromosome is important in pedigree analysis. Theoretically, obtaining these data using chromosome microdissection and PCR, as described earlier, is possible. Determining the coupling phase would require analysis of multiple PCR products using a *single* chromosome fragment as template. Alleles can be identified by either restriction fragment analysis or direct DNA sequencing of the amplification products of the target sequences. The major problem in this strategy is the presence of minute amounts of contaminating DNA on slides, on glassware, and in reagents that might obscure the coupling phase of genes in a single dissected fragment.

Recombinant DNA Libraries Generated from Microdissected Chromosome Fragments

Recombinant genomic DNA libraries may be derived from the whole genome, specific chromosomes, and chromosome regions or bands in decreasing order of genetic complexity. Recombinant libraries can be generated from microdissected chromosomal DNA either by direct microcloning or by PCR based methods as described in Chapter 4. Although methods in current use yield libraries that represent only a small fraction of unique sequences within the dissected region, these recombinant clones can be valuable as entry sites into poorly defined regions or as probes to identify and localize clones from genomic libraries that contain large DNA inserts. These clones can be used in genome mapping studies or to elucidate a number of structural and biological problems relevant to specific chromosome regions. The potential uses of microdissected chromosome libraries are described here.

Genetic Analysis of Specialized Chromosome Structures

Specialized chromosome regions are particularly amenable to microdissection and subsequent cloning. These regions include centromeres and telomeres and the regions adjacent to these structures, short arms of acrocentric chromosomes, translocations, and fragile sites. The structural basis of light and dark Giemsa banding as well as of the banding patterns generated by newer staining methods can be investigated also.

Applications in Genomic Sequencing Projects

Chromosome microdissection is likely to be used increasingly in genome sequencing projects to "fill in" specific chromosome regions that are mapped poorly. The initial phase in the development of a physical map (at the base sequence level) of human chromosomes will be to establish landmarks at approximately 100,000-base pair intervals along the genome. The landmarks will be tracts of DNA sequences, 200–500 base pairs in length called sequence-tagged sites (STSs) (4). STSs can be retrieved easily by PCR using unique oligodeoxynucleotide primers at opposite ends of the sequence tract. Thus, microdissection of a poorly defined chromosome region, followed by cloning and sequencing of unique sequence clones, rapidly could generate a pool of potential STSs from a defined region. Chromosome microdissection may be useful in constructing "contig maps" (a series of clones representing overlapping regions of the genome) as well. In this methodology, microdissected DNA is either cloned or amplified by PCR, then used to probe genomic libraries that contain larger DNA inserts, for example, cosmid or yeast artificial chromosome (YAC) libraries, under conditions that suppress repeat sequence DNA (5, 6). Mapping expressed genes to specific chromosome regions also can be accomplished using similar probes in hybridization experiments with appropriate cDNA libraries (6).

Characterization of Disease-Related Genetic Loci

A large and rapidly growing number of inheritable diseases and traits has been mapped to chromosome regions using genetic linkage methods. The isolation of the causative gene(s) can be expedited by generating a library restricted to DNA sequences from that region. Horsthemke, Lüdecke, and their colleagues as well as others investigators (5–20) have developed several DNA libraries from chromosome regions linked to inheritable diseases. These libraries are listed in Table 5.1. Once clones that are within one million base pairs of the

Table 5.1 Recombinant DNA Libraries from Microdissected Chromosomal Regions Associated with Inheritable Disorders

Disease	Chromosome region	Reference
Adenomatous polyposis coli	5q21	(7, 8)
Beckman–Wiedemann syndrome	11p15.5–pter	(9)
Cystic fibrosis	7q	(10)
Down's syndrome	21q	(11)
Fragile X syndrome	Xq27–Xq28	(12–14)
Langer–Giedion syndrome	8q24.1	(15–17)
Meningioma	22q 12–q13	(16)
Neurofibromatosis—Type II	22	(18)
Prader–Willi syndrome	15q11.1–12	(16, 19)
Tricho-rhino-phalangeal syndrome—Type I	8q23.3–924.11	(20)
Waardenburg syndrome—Type I	2q33–qter	(20)
Wilms tumor/anoridia	11p13	(5, 16)

disease locus are isolated, other molecular biological techniques may be used (e.g., chromosome walking and jumping) to isolate the gene of interest.

Study of Chromosome Abnormalities in Cancer Cells

A large number of nonrandom chromosome abnormalities are associated with specific cancer cell types (21). Translocation sites may be microdissected directly and cloned to expedite the isolation of the causative chimeric gene. Also, regions corresponding to a deletion site may be dissected from the appropriate normal chromosome. Clones then may be used as probes in *in situ* hybridization experiments to delineate whether the isolated sequences fall within the region encompassed by the deletion (22, 23). Regions of gene amplification (homogenous staining regions, double minute chromosomes) are relatively large structures that can be microdissected easily (24) for cloning and subsequent molecular analysis.

Gene Transfer Using Chromosome Fragments

Gene transfer into tissue culture cells or embryos has become an exceptionally useful method to analyze regulatory and coding regions of genes functionally. These methods are also used to induce phenotypic changes in cells and whole organisms. The preferred vehicle for gene transfer and expression is a cloned gene (usually cDNA) with appropriate flanking regulatory regions (promoters, enhancers) (25). Alternatively, whole genomic DNA has been transferred, usually with dominant selectable markers. Using microdissected fragments from specific chromosome regions in gene transfer experiments would have several potential advantages.

1. The gene or locus of interest does not have to be identified, isolated, and cloned prior to gene transfer. This ability would be particularly advantageous with respect to disease-related loci that have been

mapped genetically to a chromosome band by linkage analysis but are not further characterized.

2. The use of chromosome fragment-mediated gene transfer into embryos may be a useful tool in generating animal disease models and in confirming genetic linkage studies that map disease-causing genes to specific regions. The disease must behave as a positive dominant-acting genetic trait and not be associated with a structural deletion to be transferred to an appropriate animal.

3. Transfer of large contiguous DNA stretches from microdissected chromosome fragments allows a gene to be presented "in context," that is, with its own promoters, enhancers, control regions, and introns. All these DNA sequences may be important in appropriate cell type-specific expression of the transgene (26, 27). Further, a specific insert size may be achieved by chromosomal fragment transfer that allows "position-independent" (i.e., not influenced by the integration site in the host genome) expression of the transferred gene(s) (28).

Obvious theoretical disadvantages to chromosome fragment-mediated gene transfer are that only low (single) copy gene transfer is technically feasible. A presumed limit exists on the size of DNA accepted by the recipient cell and the methodology is labor intensive. Further, although a large stretch of DNA is microdissected (provided chromosomes have been acid-fixed and aged minimally), physical manipulation and microinjection undoubtedly results in shearing of DNA into smaller fragments. Whether multiple linear genomic DNA fragments will be more susceptible to point mutations and structural rearrangements than plasmids that are transferred is not clear. Although this method may result in position-independent integration, transferred DNA would be subject to DNA methylation controlled by genotype-specific modifiers, as occurs with other transgenes (25).

Richa and Lo (29) have microdissected and transferred human centromeric DNA successfully to fertilized mouse ova. Human centromeric DNA was detected in blastocysts, 15-day embryos, and a mouse derived from an injected ovum. The efficiency of gene transfer appeared low and all positive animals demonstrated mosaicism, probably resulting from transfer of DNA in a particulate form (i.e., there may have been delay in DNA integration). These results have not yet been confirmed by other investigators.

Microinjection of an insoluble or partially solubilized chromatin particle is likely to be an inefficient vehicle for gene transfer. An alternative method would be microdissecting a large number of chromosome fragments and collecting them in a small drop (e.g., microdrop under oil). DNA can be extracted using proteolytic enzymes or organic solvents. In Figure 5.2, a stage suitable for these experiments is shown. Microdissected chromosomal DNA is placed in an aqueous drop under oil on a depression slide (as described in Chapter 3). The slide is placed on the circular stage used for microinjection (Figure 5.2). Fertilized ova can be introduced directly under oil on the depression slide at a distance from the drop containing DNA. The DNA can be drawn up directly and introduced into ova held with a holding pipette (30). Alternatively, DNA can be drawn up into an injection pipette and microinjected into tissue culture cells on a culture dish placed on the stage.

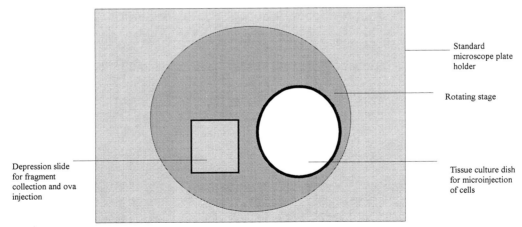

Figure 5.2 Microscope stage used for microinjection of microdissected chromosomal DNA into fertilized ova or tissue culture cells.

References

1. Cotter, F. E., Lillington, D., Hampton, G., Riddle, P., Naispuri, S., Gibbons, B., and Young, B. (1991). Gene mapping by microdissection and enzymatic amplification: Heterogeneity in leukemia-associated breakpoints on chromosome 11. *Genes Chrom. Cancer* 3, 8–15.

2. Meltzer, P., Guan, X.-Y., Burgess, A., and Trent, J. (1992). Rapid generation of region specific probes by chromosome microdissection and their application. *Nature Genetics* 1, 24–28.

3. Lengauer, C., Eckelt, A., Werth, A., Endlich, N., Ponelies, W., Lichter, P., Grulich, K. O., and Cremer, T. (1991). Painting of defined chromosome regions by *in situ* suppression hybridization of libraries from microdissected chromosomes. *Cytogenet. Cell Genet.* 56, 27–30.

4. Olson, M., Hood, L., Cantor, C., and Bolstein, D. (1989). A common language for physical mapping the human genome. *Science* 245, 1434–1435.

5. Davis, L., Senger, G., Lüdecke, H.-J., Claussen, U. Y., Horsthemke, B., Zhang, S., Metzroth, B., Hohenfellner, K., Zabel, B., and Shows, T. (1990). Somatic cell hybrid and long-range physical mapping of 11p13 microdissected genomic clones. *Proc. Natl. Acad. Sci. USA* 87, 7005–7009.

6. Yu, J., Hartz, J., Xu, Y., Gemmill, R. M., Korenberg, J., Patterson, D., and Kao, F.-T. (1992). Isolation, characterization, and regional mapping of microclones for human chromosome 21 microdissection library. *Am. J. Hum. Genet.* 52, 263–272.

7. Trautmann, U., Leuteritz, G., Senger, G., Claussen, U., and Ballhausen, W. (1991). Detection of APC region-specific signals by nonisotopic chromosomal *in situ* suppression (CISS)—Hybridization using a microdissection library as probe. *Hum. Genet.* 87, 495–497.

8. Hampton, G., Leuteritz, G., Lüdecke, H.-J., Senger, G., Trautmann, U., Thomas, H., Solomon, E., Bodmer, W., Horsthemke, B., Claussen, U., and Ballhausen, W. (1991). Characterization and mapping of microdissected genomic clones from the adenomatous polyposis coli (APC) region. *Genomics* **11**, 247–251.

9. Newsham, I., Claussen, U., Lüdecke, H.-J., Mason, M., Senger, G., Horsthemke, B., and Cavenee, W. (1991). Microdissection of chromosome band 11p15.5. *Genes Chrom. Cancer* **3**, 108–116.

10. Kaiser, R., Weber, J., Grezechik, K.-H., Edström, J. E., Driesel, A., Zengerling, S., Buchwald, M., Tsui, L. C., and Olek, K. (1987). Microdissection and microcloning of the long arm of human chromosome 7. *Mol. Biol. Rep.* **12**, 3–6.

11. Kao, F. T., and Yu, J. W. (1991). Chromosome microdissections and cloning in human genome and genetic disease analysis. *Proc. Natl. Acad. Sci. USA* **88**, 1844–1848.

12. MacKinnon, R. N., Hirst, M. C., Bell, M. V., Watson, J. E. V., Claussen, U., Lüdecke, H. J., Senger, G., Horsthemke, B., and Davies, K. E. (1990). Microdissection of the fragile X region. *Am. J. Hum. Genet.* **47**, 181–187.

13. Hirst, M. D., Roche, A., Flint, T. J., MacKinnon, R. N., Basset, J. H. D., Nakahou, Y., Watson, J. E. V., Bell, M. V., Patterson, M. N., Boyd, Y., Thomas, N. S. T., Knight, S. J. L., Warren, S. T., Hors-Cayla, M., Schmidt, M., and Davies, K. E. (1991). Linear order of new and established markers around the fragile size at x927.3. *Genomics* **10**, 243–249.

14. Djabali, M., Nguyen, C., Biunno, I., Oostra, B. A., Mattei, M. G., Ikeda, S.-E., and Jordan, B. R. (1991). Laser microdissection of the fragile X region: Identification of cosmid clones and conserved sequences in this region. *Genomics* **10**, 1053–1060.

15. Lüdecke, H.-J., Senger, G., Claussen, U., and Horsthemke, B. (1989). Cloning defined regions of the human genome by micro-

dissection of banded chromosomes and enzymatic amplification. *Nature* **238,** 348–350.

16. Lüdecke, H.-J., Senger, G., Claussen, U., and Horsthemke, B. (1990). Construction and characterization of band-specific DNA libraries. *Hum. Genet.* **84,** 512–516.

17. Lüdecke, H.-J., Johnson, C., Wagner, M., Wells, D., Turleau, C., Tommerup, N., Latos-Bielenska, A., Sandig, K. R., Meinecke, P., Zabel, B., and Horsthemke, B. (1991). Molecular definition of the shortest region of deletion overlap in the Langer–Giedion syndrome. *Am. J. Hum. Genet.* **49,** 1197–1206.

18. Fiedler, W., Claussen, U., Lüdecke, H.-J., Senger, G., Horsthemke, B., van Kessel, A. G., Goertzen, W., and Fahsold, R. (1991). New markers for the neurofibromatosis-2 region generated by microdissection of chromosome 22. *Genomics* **10,** 786–791.

19. Buiting, K., Neumann, M., Lüdecke, H.-J., Senger, G., Claussen, U., Antich, J., Passarge, E., and Horsthemke, B. (1990). Microdissection of the Prader–Willi syndrome chromosome region and identification of potential gene sequences. *Genomics* **6,** 521–527.

20. Hirota, T., Tsukamoto, K., Deng, H. Y., Yoshiura, K.-I., Ohta, T., Tohma, T., Kibe, T., Harada, N., Jinno, Y., and Nidawa, N. (1992). Microdissection of human chromosome regions 8q23.3-q24.11 and 2q33-qter: Construction of DNA libraries and isolation of their clones. *Genomics* **14,** 349–354.

21. Mitelman, F. (1988). "Catalog of Chromosome Aberrations in Cancer." A. R. Liss, New York.

22. Martinsson, T., Weith, A., Cziepluch, C., and Schwab, M. (1989). Chromosome 1 deletions in human neuroblastomas. *Genes Chrom. Cancer* **1,** 67–78.

23. Guan, X-Y., Meltzer, P., Cao, J., and Trent, J. (1992). Rapid generation of region-specific genomic clones by chromosome mi-

crodissection: Isolation of DNA from a region frequently deleted in malignant melanomas. *Genomics* **14,** 680–684.

24. Sognier, M. A., McCombs, J. L., Brown, D. B., Lynch, G. B., Eberle, R. L., and Belli, J. A. (1992). Use of chromosome microdissection and the polymerase chain reaction to characterize double minutes and homogeneously staining regions. *Proc. Am. Assoc. Cancer Res.* **33,** 360(A).

25. Raju Kucherlapate (ed.) (1986). "Gene Transfer." Plenum, New York.

26. Grosveld, F., Blom von Assendelft, G., Greaves, D., and Kollias, G. (1987). Position-independent, high-level expression of human β-globin gene in transgenic mice. *Cell* **51,** 975–985.

27. Palmiter, R., Sandgren, E. P., Avarbock, M., Allen, D. D., and Brenster, R. (1991). Heterologous introns can enhance expression of transgenes in mice. *Proc. Natl. Acad. Sci. USA* **88,** 478–482.

28. Allen, N., Norris, M., and Surani, M. A. (1990). Epigenetic control of transgene expression and imprinting by genotype-specific modifiers. *Cell* **61,** 853–861.

29. Richa, J., and Lo, C. (1989). Introduction of human DNA into mouse eggs by injection of dissected chromosome fragments. *Science* **245,** 175–177.

30. Hogan, B., Costantini, F., and Lacy, E. (1986). "Manipulating the Mouse Embryo." Cold Spring Harbor Laboratory Press, Cold Spring Harbor, New York.

Appendix

List of Suppliers and Addresses

American Type Culture Collection (ATCC)
12301 Parklawn Drive
Rockville, Maryland 20852
(800) 638-6597

A-M Systems, Inc.
1220 75th Street S.W.
Everett, Washington 98203
(800) 426-1306

Bethesda Research Laboratories (BRL)
P.O. Box 6009
Gaithersburg, Maryland 20887
(800) 638-8992

Biomedical Specialties
P.O. Box 1687
Santa Monica, California 90406
(310) 454-1995

Bio-Rad Laboratories
1414 Harbour Way South
Richmond, California 94804
(800) 227-3259 (West)
(800) 645-3227 (East)

BIOS Laboratories
291 Whitney Avenue
New Haven, Connecticut 06511
(800) 678-9487

Boehringer-Mannheim Biochemicals
9115 Hague Road
Indianapolis, Indiana 46250
(317) 576-2771

Carl Zeiss, Inc.
One Zeiss Drive
Thornwood, New York 10594
(914) 747-1800

Corning Glass Works
Corning, New York 14831
(607) 974-4667

COSTAR
One Alewife Center
Cambridge, Massachusetts 02140
(800) 492-1110

Dage-MTI, Inc.
710 N. Roeske Avenue
Michigan City, Indiana 46360
(219) 872-5514

Difco Laboratories
P.O. Box 1058
Detriot, Michigan 48232
(313) 961-0800

Diamond General Corporation
3810 Varsity Drive
Ann Arbor, Michigan 48108
(313) 973-0179

Eppendorf Corporation
45635 Northport Loop East
Fremont, California 94538
(415) 659-0181

Ericomp, Inc.
10055 Barnes Canyon Road, Suite G
San Diego, California 92121
(619) 457-1888

Falcon Labware
(see Fisher Scientific)

Fisher Scientific
711 Forbes Avenue
Pittsburgh, Pennsylvania 15219
(412) 562-8300

GenoSys (Genetic Design)
7505 S. Main Street Suit 270
Houston, Texas 77030
(713) 795-4686

GIBCO BRL
8400 Hellerman Court
Gaithersburg, Maryland 20877
(301) 840-8000

Graticules, Ltd.
Ton Bridge, Kent
England

Imagenetics, Inc.
31 New York Avenue
Framingham, Massachusetts 01701
(508) 872-3113

Kinetic Systems, Inc.
20 Arboretum Road
P.O. Box 414
Roslindale, Massachusetts 02131
(617) 522-8700

Medical Systems Corporation
One Plaza Road
Greenvale, New York 11548
(800) 654-5406

Narishige USA, Inc.
One Plaza Road
Greenvale, New York 11548
(800) 445-7914

New England Biolabs, Inc.
32 Tazor Road
Beverly, Massachusetts 01915
(800) 632-5227

Nikon, Inc.
Instrument Group
Biomedical Instrument Department
623 Stewart Avenue
Garden City, New York 11530
(516) 222-0200

Oncogene Science Inc.
106 Charles Lindbergh
Uniondale, New York 11553-3649
(800) 662-2616

ONCOR, Inc.
209 Perry Parkway
Gaithersburg, Maryland 20877
(800) 776-6267

Perkin Elmer Corporation
761 Main Avenue
Norwalk, Connecticut 06859-0156
(800) 762-4002

Pharmacia LKB Biotechnology, Inc.
800 Centennial Avenue
Piscataway, New Jersey 08854
(800) 558-7110

Qiagen, Inc.
9259 Eton Avenue
Chatsworth, California 91311
(800) 426-8157

Schleicher & Schuell, Inc. (S & S)
Keene, New Hampshire 03431
(800) 245-4024

Stoelting Co.
620 Wheat Lane
Wood Dale, Illinois 60191
(708) 860-9700

Stratagene
11099 North Torrey Pines Road
La Jolla, California 92037
(800) 548-1113

Sutter Instruments Co.
40 Leveroni Ct.
Novato, California, 94949
(415) 883-0128

Technical Products International Inc.
5918 Evergreen
St. Louis, Missouri 63134
(800) 729-4421

US Biochemical Corporation
P.O. Box 22400
Cleveland, Ohio 44122
(800) 321-9322

VWR Scientific
CN 1380
Piscataway, New Jersey 08854
(201) 756-8030

Vector Laboratories, Inc.
30 Ingold Road
Burlingham, California 94010
(415) 697-3600

Video Scope International, Ltd.
Dulles International Airport
Washington, DC 20041-7150
(703) 437-5534

Wild Leitz USA; Inc.
MQM Division
24 Link Drive
Rockleigh, New Jersey 07647-9987
(201) 767-1100

World Precision Instruments (WPI), Inc.
175 Sarasota Center Boulevard
Sarasota, Florida 34240-9258
(813) 371-1003

Glossary

alkaline phosphatase Enzyme that is used to remove the 3' phosphate group from DNA strands to prevent self-ligation during cloning. This enzyme also can be conjugated to biotin to function as part of the detection system for biotinylated probes. Reaction of the enzyme with the substrate BCIP (5-bromo-4-chloro-3-indolyl phosphate) generates a colored product.

allele One of several alternative forms of a DNA sequence occupying a given locus on the chromosome. A single allele for each locus is inherited separately from each parent.

allele-specific oligonucleotide probes DNA probes ranging in size from 15 to 25 nucleotide base pairs, exactly complementary to a normal gene sequence or a sequence with a point mutation.

***Alu* primer** Oligonucleotide complementary to sequences in the region of the interspersed *Alu* repeat sequences. These primers allow the amplification of genomic sequences between these repeats using the polymerase chain reaction technique.

amplification Production of additional copies of a chromosomal sequence, usually in tandem array.

anneal To associate two complementary single-stranded nucleic acid strands through hydrogen bond formation.

autoradiography Technique that uses X-ray film to visualize radioactively labeled molecules or fragments of molecules; used in analyzing length and number of DNA fragments after they are separated by gel electrophoresis.

autosome Chromosome not involved in sex determination. The diploid human genome consists of 46 chromosomes, 22 pairs of autosomes and 1 pair of sex chromosomes (the X and Y chromosomes).

bacteriophage (phage) Virus that infects bacteria.

base pair (bp) Hydrogen-bonded complex of A and T (or U in RNA) or of C and G between two complementary nucleic acid strands.

biotinylation of nucleic acids Nonradioactive method of labeling nucleic acid probes using nick translation to incorporate biotin-derivatized nucleotides.

cDNA Single-stranded DNA complementary to an mRNA molecule, synthesized from it by reverse transcription.

cDNA clone Double-stranded DNA sequence derived from an mRNA; contains only exon sequences.

cDNA libraries Collection of cDNA clones derived from a particular cell or tissue, propagated in cloning vectors.

centimorgans (cM) Named after the American geneticist Thomas Morgan. Two markers are said to be 1 cM apart if they are separated by recombination 1% of the time. A genetic distance of 1 cM is roughly equal to a physical distance of one million base pairs (1 Mbp).

centromeric DNA DNA in the constricted region of a chromosome including the attachment site for the mitotic spindle.

chromatin Complex of DNA and protein in interphase nuclei.

chromosomal bands Variation in the staining of DNA due to differential binding of dyes along the length of each DNA molecule.

chromosome Self-replicating genetic structure of cells that contains the cellular DNA that bears in its nucleotide sequence the linear array of genes.

chromosome fragment Small portion of chromosomal DNA that represents a band (stained or unstained) on the chromosome obtained by means of physical excision using microdissection.

clone Large number of identical cells or molecules derived from a single ancestor.

cloning Mechanism by which one attempts to isolate a unique nucleic acid sequence.

cloning vector Any plasmid or phage into which a foreign DNA may be inserted and propagated in a heterologous host (such as bacterial, viral, or eukaryotic cells).

colony Visible growth of microorganisms or cells on solid microbiological medium.

competent Particular condition of cells, such as *E. coli*, following chemical treatment that makes the cell envelope permeable to exogenous DNA molecules.

contig map Map depicting the relative order of a linked library of small overlapping genes representing a complete chromosomal segment.

cosmids Artificially constructed cloning vectors in which phage λ cos sites have been inserted. As a result, the plasmid DNA can be packaged *in vitro* in the phage coat for infection into *E. coli*. This permits the cloning of larger DNA fragments (up to 45 kb) that can be introduced into bacterial hosts.

Cot1 DNA Contains highly repetitive DNA derived from high molecular weight DNA that is isolated from human placentas and blended to homogeneity. Sheared DNA is denatured by heating and renatured to a Cot of approximately 1 [Cot: concentration (moles nucleotides/liter) × time (seconds)], treated with S1 nuclease, extracted with phenol chloroform, ethanol precipitated, and resuspended in 10 m*M* Tris, pH 7.5, 1 m*M* EDTA (TE).

denaturation Causing strand separation of double-stranded nucleic acid by heating or treating with alkali to enable hybridization of the resulting single-stranded species.

diploid Having a full set of genetic material consisting of paired chromosomes, one from each parental set. Most animal cells except the gametes have a diploid set of chromosomes.

DNA polymerase (e.g., *E. coli* DNA) Polymerase enzyme that contains an excision repair function that is taken advantage of in the nick translation reaction to incorporate labeled nucleotides into a DNA probe.

DNA replication Use of existing DNA as a template for the synthesis of new DNA strands. In humans and other eukaryotes, replication occurs in the nucleus.

DNase Endonuclease that degrades unprotected DNA to mono- and dinucleotides.

electrophoresis Method used to separate charged molecules that migrate in response to the application of an electrical field. In gel electrophoresis, heterogenously sized molecules in a sample are drawn through an inert matrix such as agarose and separate during migration according to size.

3′ end End of a piece of DNA (or RNA) that contains the hydroxyl group attached to the 3′ carbon of the corresponding base.

5′ end End of a piece of DNA (or RNA) that contains the phosphate group attached to the 5′ carbon of the corresponding base.

endonucleases Nucleases that cleave bonds within a nucleic acid; they may be specific for RNA or DNA.

eukaryotes Organisms with cells that have limiting membranes around nuclear material.

exon Actual coding sequence in DNA, which is interrupted by noncoding sequences (introns).

expression vector Vector in which the inserted DNA molecule is placed near a promoter, allowing mRNA and protein expression of the cloned sequence in a heterologous host.

filter hybridization Hybridization performed by incubating a denatured DNA preparation immobilized on a nitrocellulose filter with a solution of radioactively labeled RNA or DNA.

filter sterilization Sterilization of liquid medium or medium component, as an alternative to autoclaving, by suction through a selective membrane in a sterile container.

flow cytometry Analysis of biological material by detection of the light-absorbing or fluorescing properties of cells or subcellular fractions (i.e., chromosomes) passing in a narrow stream through a laser beam. An absorbance or fluorescence profile of the sample is produced. Flow cytometry also has been used to analyze and/or separate chromosomes on the basis of their DNA content (flow karyotyping).

fluorescence *in situ* hybridization (FISH) Physical mapping approach that uses fluorescein or other fluorescing molecules to detect hybridization probes with metaphase chromosomes and with less-condensed somatic interphase chromatin.

gene Fundamental physical and functional unit of heredity. A gene is an ordered sequence of nucleotides located in a particular position on a particular chromosome and encodes a specific functional product.

gene mapping Determining the relative positions of genes on a DNA molecule and of the distance, in linkage units or physical units, between them.

genetic linkage analysis Analysis of cosegregation of polymorphic DNA sequences in the population with disease phenotypes, thus establishing "linkage" between the genetic polymorphism (and its chromosmal location) and the disease-associated gene.

genetic marker Trait whose inheritance can be followed within a family.

genome All the genetic material in the chromosomes of a particular organism; its size generally is given as its total number of base pairs.

genomic (chromosomal) DNA clones Genomic DNA sequences (containing introns and exons) inserted in a cloning vector.

genotype Genetic constitution of an organism.

haploid genome Single set of chromosomes containing one of each of the autosomes and one sex chromosome (i.e., half the full set of genetic material).

heterozygosity Presence of different alleles at one or more loci on homologous chromosomes.

homogeneously staining region Extended region of a chromosome that has a homogeneous banding pattern and is thought to be a site of gene amplification.

homology Degree of complementarity between two nucleic acid strands or two coding sequences.

host Organism that has been infected or transformed with a virus or plasmid and that is intended to be a growth or study chamber for the newly introduced DNA on the viral or plasmid vector.

hybridization Pairing of complementary RNA and DNA strands to give an RNA–DNA or DNA–DNA hybrid through the formation of hydrogen bonds between complementary nucleotides.

***in situ* hybridization** Use of a DNA or RNA probe to detect the presence of a complementary DNA sequence of a particular gene(s) within bacterial or eukaryotic cells.

interphase Period in the cell cycle when DNA is replicated in the nucleus, followed by mitosis.

interspersed repeats Families of DNA sequences that are present in a single copy in multiple locations throughout the genome (e.g., *Alu* and L1 repeat sequences).

intron Noncoding DNA base sequences interrupting the protein-coding sequences (exons) in a gene. These are removed during the processing of messenger RNA so mature messenger RNA contains only coding exon sequences.

karyotype Photomicrograph of an individual's chromosomes arranged in a standard format showing the number, size, and shape of each chromosome type; used in low resolution physical mapping to correlate gross chromosome abnormalities with the characteristics of specific disease.

karyotyping Visualizing the entire chromosome complement of a cell or species during mitosis.

kilobase (kb) 1000 base pairs of DNA or 1000 bases of RNA.

LB (Luria–Bertani) Complex rich medium for culturing bacteria.

library Set of an unordered collection of cloned DNA fragments, all derived from the same cell, tissue, or organism.

ligases Enzymes that promote formation of phophodiester bonds between the adjacent nucleotides.

linkage Tendency of genes to be inherited together as a result of their location on the same chromosome; measured by percentage of recombination between loci.

locus (plural, loci) Position on a chromosome at which the gene for a particular trait resides.

megabase (mb) Unit of length for DNA fragments equal to one million base pairs of nucleotides.

melting Denaturation (of DNA).

melting temperature (t_m) Midpoint of the temperature range over which DNA is denatured.

messenger RNA (mRNA) RNA molecule copied from the DNA template containing a gene that will encode a protein.

metaphase chromosomes Condensed chromosomes at metaphase attached to the mitotic spindle through centromeres.

microcentrifuge Small tabletop centrifuge, with a radius of 40–50 mm and rotor holes for microcentrifuge tubes, which is capable of speeds of up to 14,000 rpm.

microdissection Method to physically cut and collect chromosomal fragments by means of very fine needles and a micromanipulator.

microsyringe (micropipette) Very fine pipettes with a tip diameter less than 1 μm that are fashioned to cut and collect very small fragments of chromosomal DNA. Micropipettes are also used to inject nanoliter volumes into cells and ova *in vitro*.

nick Absence in duplex DNA and RNA of a phosphodiester bond between two adjacent nucleotides on one strand.

nick translation Ability of DNA polymerase I to use a nick as a starting point from which one strand of a duplex DNA can be degraded and replaced by synthesis of new material. The process is used to introduce radioactively labeled nucleotides into the DNA *in vitro*.

nitrocellulose Transfer membrane on which nucleic acids or proteins are immobilized by blotting.

Northern blotting Technique for transferring RNA from an agarose gel to a nitrocellulose filter on which it can be hybridized to a complementary probe.

nucleosome Basic structural unit of chromatin, consisting of approximately 160 bp of duplex DNA associated with an octamer of histone proteins.

nylon membrane Transfer medium for blotting DNA or protein purported to be more durable (can be used for multiple probings) and versatile than nitrocellulose.

oligonucleotides Short series of nucleotide base pairs used as primers or probes.

oncogene Genetic sequence involved in the control of growth and differentiation; disruption of these sequences may be associated with malignant transformation. Cellular oncogenes are those that have been transduced from eukaryotic cells by a retrovirus.

phenotype Appearance or any other characteristic of an organism resulting from interaction of its genetic constitution with the environment.

plasmid Autonomous self-replicating extrachromosomal circular DNA molecule.

plate (1) Petri dish. (2) To spread cells onto solid nutrient medium in a petri dish.

polymerase chain reaction (PCR) Technique by which target DNA sequences may be replicated selectively (amplified) *in vitro*. DNA primers flanking a site of interest (which hybridize to opposite strands of the duplex) are added to the denatured target, followed by the addition of DNA polymerase, which uses the primer to extend the DNA sequence of the target. After multiple cycles of denaturation, primer binding, and DNA synthesis, a target sequence may be replicated 10^6-fold *in vitro*.

polymorphism Simultaneous occurrence in the population of genomes showing allelic variations, that is, variation in the amount or sequence of DNA between two individuals.

primer Short sequence (often of RNA) that is paired with one strand of DNA to provide a free 3'-OH end at which a DNA polymerase starts synthesis of a deoxyribonucleotide chain, using the full-length strand as a template.

probe (1) Piece of labeled DNA or RNA used to locate immobilized sequences on a blot by hybridizing under optimal conditions of salt and temperature. (2) To hybridize a probe to a blot or in solution.

promoter Region of DNA immediately upstream of the start site of transcription of a gene; involved in binding RNA polymerase to initiate transcription.

rare base cutters Restriction enzymes that have an 8-bp recognition sequence that will occur only rarely (every 65,000 bp) in the genome. These enzymes cleave DNA to create very long fragments.

restriction endonuclease Enzyme that recognizes short DNA sequences and cleaves the DNA at these target sites.

restriction enzyme See restriction endonuclease.

restriction fragment length polymorphism (RFLP) Variation between individuals in DNA fragment sizes cut by specific restriction enzymes; polymorphic sequences that result in RFLPs are used as markers on both physical maps and genetic linkage maps. RFLPs usually are caused by a mutation at a cutting site.

restriction map Linear array of sites on DNA cleaved by various restriction enzymes.

restriction site Recognition sequence for a restriction enzyme. May refer specifically to the exact point of cleavage between nucleotides within the sequence.

RNases Endonucleases and exonucleases involved in the processing or degradation of RNA molecules.

sequence tagged site (STS) Short stretch of genomic sequence that can be detected by the polymerase chain reaction and mapped to a particular point in the genome where the STS then serves as a landmark.

somatic cell Any cell of an organism with the exception of germ cells.

Southern blotting analysis Procedure for transferring DNA from an agarose gel to a filter on which it can be hybridized with complementary mRNA molecules.

sticky end Complementary single strands of DNA that protrude from opposite ends of a duplex or from ends of different duplex molecules; generated by staggered cutting of restriction endonuclease target sequences.

stringency Conditions of the hybridization reaction that determine the stability of the duplex; 75% stringency implies that 25% of the bp of the two strands may be mismatched, but a duplex will still form; 100% stringency implies that two strands must be perfectly complementary for a duplex to form.

***Taq* DNA polymerase** Thermostable DNA polymerase isolated from *Thermus aquaticus*; can withstand repeated exposure to high temperatures (94–95°C). *Taq* has been used in polymerase chain reactions.

template Nucleic acid sequence (usually DNA) used as a guide to direct the synthesis of a complementary strand of nucleotides linked by phosphodiester bonds (as in DNA replication or transcription).

thermal cycler (temperature cycler) Instrument designed specifically to automate the polymerase chain reaction (PCR). Because it is microprocessor controlled, it can be programmed accurately to perform rapid temperature changes and incubations required for the PCR process. Multiple samples can be amplified simultaneously using the same program.

transcription Synthesis of an RNA copy from a sequence of template DNA.

transfection Transfer of foreign DNA into heterologous cells.

transformation Process by which microorganisms accept and incorporate exogenously added DNA. Transformation occurs naturally among microbes or can be induced artificially using competent cells.

translocation Rearrangement in which part of a chromosome is detached by breakage and becomes reattached to another chromosome.

unique sequence DNA sequence that encodes a structural gene that has a functional protein product.

vector See cloning vector.

yeast artificial chromosome (YAC) Cloning vector capable of accommodating larger pieces of genomic DNA (average insert is in excess of 500 kb); constructed from the telomeric, centromeric, and replication origin sequences needed for replication in yeast cells.

Some of the definitions used in this glossary were adapted from "The Human Genome Program Report 1991–1992" published by The United States Department of Energy.

Index

Actinomycin D, 26
Agarose
    gel analysis, 79, 90, 104
Alkaline phosphatase, 93
Alleles, 133
Alu primers, 101
    PCR, 95
    repeats, 101
    structure, 101
Alu sequences, 10
    SINES, 10
Amplification
    FISH signal, 110
    PCR, 85, 89, 93, 100, 101
Anneal, 102–103
Antibiotics
    ampicillin, 91
    penicillin, 30–31
    streptomycin, 30–31
Antifading agent, 119
Aqueous mount, 119
Avidin
    antibody, 117
    FITC-labeled, 117

Bacteriophage, 81
Biosmap blots, 109
Biotinylated DNA, 111–112, 114
    detection of, 116
    probe, 116
    separation, 112

Blood
    cells, 32
    heparinized tube, 32
    microculture method, 32
Bromodeoxyuridine, 10

Cameras, 42, 116
Cancer
    chromosome abnormality, 136
    translocations, 132, 134, 135
C-bands, 7, 10
    satellite DNA, 10
CDNA, 136
CDNA clone, 136
CDNA library, 97, 109, 135
Cells
    blood cells, 32
    lymphocytes, 32
    tissue culture, 34, 35
Centromeres, 11–13, 134
Chromatid, 5
Chromatin, 2
    fiber, 2, 4, 12
    loops, 4
Chromatin particle,
    microinjection of, 137–138
Chromomycin A3, 6
Chromosomal band resolution, 8–9, 25
Chromosomal DNA
    cloning of, 77

degradation, 84
methods of cloning, 73, 85, 95
precautions of, 97
PCR of, 95
primers, 99, 101
summary protocol, 96
Chromosomes, 1, 14
 aging, 25, 29
 cell cycle, 28
 degradation, 25
 depurination, 28, 85
 DNA fixation, 25, 28, 85
 Drosophila polytene, 11
 preparation from peripheral blood
   lymphocytes, 32
  nicking, 28
  nomenclature, 11
  preparation, 26
   from Dipteran salivary glands of
    Drosophila melanogaster, 36
   from monolayer culture cell lines, 34
   from monolayer cells on coverslips,
    35
  scaffold, 4
  staining, 29–30
  storing, 29
Chromosome bands, 2, 5–7
 C-bands, 10
 cell cycle and, 6
 chromatin organization, 6
 DNA replication and, 10
 G bands, 10
 gene expression and, 10
 genome organization, 6
 Giemsa dark bands, 6
 metaphase, 6
 model of, 7
 polytene chromosomes, 10
 prophase, 6
 R bands, 10

Chromosome fragments
 collection, 59–60
 dissection, 59–60
Chromosome organization, 1
Chromosome painting, 120
Chromosome R banding
 DAPI, 119
 Hoechst dye, 118
Chromosome regions, 11
 centromeres, 11–13
 telomeres, 11, 13–14
Chromosome staining, 118
Chromosome Work Station, 42, 58
Cloning efficiency
 estimation of, 107–108
Cloning
 ligation and PCR, 87–88, 91–92
 summary of, 86
Colcemid, 26, 30, 32, 35
Colony
 bacterial, 91
Competent bacteria, 91, 93
Contig map, 134
Cosmid, 95, 135
Cot1 DNA, 103
 suppression hybridization, 116

DABCO, *see* Antifading agent
DAPI, 119
Denaturation, 25, 28, 85
Depression slide, 82, 98
Direct cloning, 79
 phage, 81
 summary of, 80
Dissection chamber, 49–50
Dissection micropipette, 55, 81
DNA
 Alu sequences, 10, 95, 101

cloned, 77, 107
contamination, 15, 133
insert size, 104, 107
LINES, 10
loops, 2, 4
recombinant DNA
  clones 77
  complexity, 107–109
  genomic localization, 109
  library, 78, 95
  methods, 77
  molecules, 77
  repeat sequences, 105
  structural genes, 109
  SINES, 10
  unique sequences, 106
repeats, 10, 105
replication, 2
DNase, 25, 114

Electrophoresis, 90
Endonucleases, 82
Ethidium bromide, 27, 90, 104

Ficoll–Paque, 30
Filter, 31, 87
Filter hybridization, 87
Filter sterilization, 37
FISH, 119, 132
  micro-FISH, 132

Gel electrophoresis, 90
Gene
  locus, 136
  mapping, 131
  transfer, 131, 136

Gene amplification
  double minutes, 136
  homogeneous staining region (HSR), 136
Genetic disease, 15
Genetic linkage, 135
Genome, 6
  bands, 6
Genomic DNA, 88
Genomic library, 97, 133
  types of, 135
Giemsa stain, 31
  solid staining and GTG banding of metaphase chromosomes, 36–37
GTG banding, 25, 30

Helical coiling, 2, 4
  chromatin fiber, 4
Heparin, 31–32
Heterochromatin, 7
Histone proteins, 2, 3–5
  acylation, 5
  methylation, 5
  phosphorylation, 5
Hoechst, 118
Human genome project, 134
Hybridization
  detection of, 118
  fluorescent, 110, 120
  in situ, 115
  signal, 117
Hypotonic treatment
  procedure, 27–28

In situ hybridization, 104, 109
  probe preparation for, 111–112
  protocol, 114
  summary of, 113
Introns, 137

Karyotype, 6, 8
    analysis, 132
    human karyotype, 8
    normal, 8–9
Kinetochore, 12

L1 repeats, 101
    primers, 101
    structure, 101
Laser microbeam, 67
    apparatus, 67–68
    microdissection, 68–69
LB broth, 91
Ligase, 81, 88
Ligation, 83, 85
LINES, 10, 101
    base composition, 10
Linkage, 133
    genetic, 135
Linker adaptor cloning, 85, 91–92
    primers, 91
Linker DNA, 2
Lymphocytes, 27
    chromosome preparation, 32
    Ficoll–Hypaque gradient, 33

Matrix-associated regions (MAR), 4–5
MboI, 87, 91
    digestion with, 92
Metaphase chromosomes, 26
    cell cycle, 26
    preparation of, 26–27
    scaffold, 5
    spreads, 26

Methotrexate, 26
Microdissection, 1, 59–63
    apparatus, 42, 65, 67
    applications of, 131
    cancer cells and, 136
    chromosome abnormality, 136
    determining coupling phase, 133
    disease loci, 135
    genes mapped by, 135
    genetic analysis, 134
    gene transfer, 136–137
    making recombinant libraries, 133
    mapping sites, 132
    methods, 42
        laser microbeam, 67
        oil chamber, 65
        video microscope, 42
    STS and, 134
    summary, 70–73
    translocation, 132
    YAC libraries and, 135
Microforge, 56
Microgrinder, 57
Microinjection, 138
    cells in culture, 138
    mouse embryos, 137
Microinjector, 56
Micromanipulator, 49, 51
Micropipette, 52, 55
    cutting, 55
    holder, 51
    puller, 53
    tip measurement, 54
Microscope coupler, 43, 45
Microsyringe, 81
Mitotic cells, 34
    thymidine block, 34
    shake-off, 34
Mouse embryos, 137

Nanoliter drop, 14
    making of, 60
    transfer, 98
Nick, 27–28
    nick translation, 106, 112
Nitrocellulose, 87
Nuclear matrix, 4
Nucleosomes, 2

Oil chamber, 65
    microdissection and, 66
Oligonucleotide, 132
    primers, 90, 91, 97, 99, 101

Packaging extract, 83
    into phage, procedure of, 83
Paraffin oil, 81
Phage, 78, 79
Phosphate-buffered saline (PBS), 30
Physical map, 134
Phytohemagglutinin, 30
Pipette method, 24
Plaques, 83, 84
Plasmid, 85, 87
Plasmid vector cloning, 85–88
    primers, 90
Polymerase chain reaction (PCR), 1, 85, 89, 93, 100
    Alu primers, 101–102
Primers
    Alu, 10, 101
    degenerate, 101
    restriction site, 91, 99
    universal, 90, 97, 99

Probe(s), 134
Promoter, 137
Propidium iodide, 119
    counterstaining with, 118
Proteinase K, 79, 111, 116

Quinacrine mustard, 6

Random primer
    labeling DNA with, 100
Recombinant clones
    analysis of, 104–120
Repeat sequences, 103
    assay of, 105
    blur 8, 106
Replication unit, 5
Replicon, 5
Resolution mapping, 110
Restriction endonuclease, 82
    DNA characterization,
        *Bam*HI, 93, 106
        *Eco*RI, 79, 81, 85
        *Hae*III, 10
        *Hin*dIII, 81
        *Pst*I, 106
    microcloning,
        *Eco*RI, 79, 81, 85
        *Hpa*II, 85, 94
        *Mbo*I, 87, 91
        *Not*I, 99
        *Rsa*I, 85, 94
        *Sma*I, 87
        *Xho*I, 99
Restriction fragment length polymorphism (RFLP), 133

Restriction map, 133
RNase, 111, 115

Satellite DNA, 12–13
　alpha, 12
　beta, 13
　C-bands, 10
Scaffold, 4
Scaffold-associated regions (SAR), 4–5
Sequence tagged site (STS), 95, 134
Silicon oil, 82
SINES, 101
　base composition, 10
Southern blot, 78, 106
Stringency, 106
Structural genes, 109
　zooblots, 109

T4 ligase, 88
Taq DNA polymerase, 88, 90, 93, 98
Telomerase, 14, 134
Telomeres, 11, 13
　structure, 13
　　C-rich strand, 14
　　G-rich strand, 14
　　length, 13
　　species, 13
Template, 77
Thermal cycler, 88, 98
Thymidine block, 25, 34
Tissue culture cells, 34, 35
　chromosome preparation, 34–35
Topoisomerase II, 4

Transcription, 2, 4
Transfection, 91
Transgene, 137–138
Translocation, 132, 134, 136
　microdissection of, 132–133
Trypsin, 31
　EDTA, 37

Unique sequences
　assay of, 106
　estimation of, 107

Vector
　pUC 13, 91
　pUC 18, 88–89
Vibration-free table, 50
Video camera, 42
　microscope 42
Video printer, 46
Volumetric micropipette, 55, 81

XGal, 91

Yeast artifical chromosome (YAC), 95, 110, 135

Zooblots, 109